ROWAN UNIVERSITY
LIBRARY
201 MULLICA HILL RD.
GLASSBORO, NJ 08028-1701

Improving Maintainability and Reliability through Design

Improving Maintainability and Reliability through Design

by

Dr Graham Thompson

MSc, PhD, CEng, FIMechE

Professional Engineering Publishing

Professional Engineering Publishing Limited
London and Bury St Edmunds, UK

First published 1999

This publication is copyright under the Berne Convention and the International Copyright Convention. All rights reserved. Apart from any fair dealing for the purpose of private study, research, criticism, or review, as permitted under the Copyright Designs and Patents Act 1988, no part may be reproduced, stored in a retrieval system, or transmitted in any form or by any means, electronic, electrical, chemical, mechanical, photocopying, recording or otherwise, without the prior permission of the copyright owners. Unlicensed multiple copying of this publication is illegal. Inquiries should be addressed to: The Publishing Editor, Professional Engineering Publishing Limited, Northgate Avenue, Bury St Edmunds, Suffolk, IP32 6BW, UK.

© G Thompson

ISBN 1 86058 135 8

A CIP catalogue record for this book is available from the British Library.

Printed and bound in Great Britain by Antony Rowe Limited, Chippenham, Wiltshire, UK

TA
174
.T47
1999

The publishers are not responsible for any statement made in this publication. Data, discussion, and conclusions developed by the Author are for information only and are not intended for use without independent substantiating investigation on the part of the potential users. Opinions expressed are those of the Author and are not necessarily those of the Institution of Mechanical Engineers or its publishers.

Related Titles of Interest

The Reliability of Mechanical Systems	J Davidson and C Hunsley	0 85298 881 8
IMechE Engineers' Data Book	C Matthews	1 86058 175 7
An Introductory Guide to the Control of Machinery	Edited by K Foster	1 86058 068 8
Managing Enterprises – Stakeholders, Engineering, Logistics, and Achievement	Edited by D T Wright, M M Rudolph, V Hanna, D Gillingwater, and N D Burns	1 86058 066 1
Design Reuse – Engineering Design	Edited by S Sivaloganathan and T M M Shahin	1 86058 132 3

For the full range of titles published by Professional Engineering Publishing contact:

Sales Department
Professional Engineering Publishing Limited
Northgate Avenue
Bury St Edmunds
Suffolk
IP32 6BW
UK

Tel: +44 (0)1284 724384
Fax: +44 (0)1284 718692

About the Author

Dr Thompson took up a lectureship in Mechanical Engineering at UMIST from BNFL in 1978, where he had worked in the R & D Department. His research activities at UMIST are motivated by a strong interest in engineering design and design for maintainability and reliability in particular. Design principles and equipment design studies have featured in the research and include applications in the process and manufacturing industries. The research has been undertaken in collaboration with industry and has received support from government and industrial funding bodies. Other research interests include creativity and creative problem solving, a field in which he is a trained facilitator and has also published. Dr Thompson's research has resulted in the publication of over 60 papers in international journals and conferences and the award of four prizes by the Institution of Mechanical Engineers. His PhD was gained in 1983 on 'The reduction of maintenance costs through design'.

Dr Thompson is presently a Senior Lecturer at UMIST. The mainstream undergraduate subjects taught include engineering design and applied mechanics subjects and at post graduate level design, maintainability, and reliability are taught. He believes design should be an exciting subject in which students are given opportunities to express and develop their ideas. Continuing education courses are given to industry on design for maintainability and reliability and creative problem solving.

Professional engineering activities have been integrated with academic teaching and research. Consultancies have been undertaken in a variety of areas including design studies, accident investigation, and legal disputes. Active in the Institution at national and local levels, Dr Thompson is presently a Vice-Chairman of the Process Industries Board and was chairman of the N. W. Branch in 1996–97. He was editor of the Journal of Process Mechanical Engineering from 1989–1997 and is presently a member of the editorial board.

Contents

Acknowledgements		xvii
Foreword		ixx
Chapter 1	**Introduction**	1
Chapter 2	**Opportunities to influence design**	5
2.1	Engineering design	5
2.2	Design activity	5
	2.2.1 Design phases	5
	2.2.2 Definition of requirements in a specification	6
	2.2.3 Concept design	7
	2.2.4 Detail design	7
2.3	Levels of design	8
2.4	Design models	9
	2.4.1 Equipment and product design	9
	2.4.2 Large-scale projects	9
	2.4.3 Iteration in design	11
2.5	The nature of design activity	12
	2.5.1 Divergent and convergent thinking	12
	2.5.2 Divergent–convergent coupled activities	12
2.6	Feedback of experience to design	13
2.7	Summary	14
Chapter 3	**Maintainability and reliability: basic principles**	17
3.1	Introduction	17
3.2	Definitions	18
3.3	Maintainability prediction in design	19
3.4	Availability	21
3.5	Reliability and failure rate	22
	3.5.1 Reliability	22
	3.5.2 Mean failure rate	23
	3.5.3 Constant mean failure rate	24
	3.5.4 Mean time to failure, MTTF	25
3.6	Reliability modelling	26
	3.6.1 Reliability modelling and design	26

	3.6.2	Series elements	26
	3.6.3	Parallel elements: active redundancy	27
	3.6.4	Combined series and parallel	28
	3.6.5	Partial active redundancy	29
3.7		Prediction of reliability using failure rate data	30
3.8		Component count method for predicting reliability and maintainability	31
3.9		Examples of the use of reliability calculations	33
	3.9.1	Improving reliability by using parallel elements	33
	3.9.2	Alternative use of parallel elements	34
	3.9.3	Reliability as a function of time	35
	3.9.4	A component count example	35

Chapter 4 Design Review — 39

4.1		Introduction	39
4.2		Levels of design	40
4.3		A structured design review procedure	40
4.4		Design specification	41
4.5		System review	42
	4.5.1	Prior to detail design	42
	4.5.2	After equipment has been designed	43
4.6		Equipment evaluation	44
4.7		Component analysis	45
4.8		The design review team	46
	4.8.1	Role of the review team	46
	4.8.2	Composition of the review team	46
	4.8.3	Management and control	47
4.9		General discussion	48

Chapter 5 Equipment Evaluation — 51

5.1		Introduction	52
5.2		Check list	52
	5.2.1	A simple check list	52
	5.2.2	The use of check list to compare equipment	53
5.3		General discussion about equipment selection	54
	5.3.1	Costs	54
	5.3.2	Component quality	55
	5.3.3	Spares, servicing, and technology	55
	5.3.4	Operating environment	56
	5.3.5	Condition monitoring	56
	5.3.6	Equipment design features	56

	5.3.7 A qualitative systematic procedure	57
5.4	Comparative reliability analysis	57
	5.4.1 Method	57
	5.4.2 Example: Comparative reliability analysis of two process valves	57
5.5	Evaluation of design concepts	60
5.6	Systematic quantitative equipment evaluation	62
	5.6.1 Introduction	62
	5.6.2 Basic principles	63
	5.6.3 A systematic quantitative method	63
	5.6.4 Comments on the method	65
	5.6.5 Device performance index	66
	5.6.6 Non-linear utility functions	67
	5.6.7 Illustrative example of a device performance index calculation	67
5.7	Systems analysis using *DPI*	70
5.8	Equipment handling fluids	71
	5.8.1 Method	71
	5.8.2 Brief case study: Ball plug valve analysis	72
5.9	Case study: an evaluation of a machine to assemble battery components	73
	5.9.1 Introduction	73
	5.9.2 Assessment criteria	73
	5.9.3 Value judgements	74
	5.9.4 Machine analysis	75
	5.9.5 Device performance index	77
	5.9.6 Decisions	78

Chapter 6	**System Evaluation: Parameter Profile Analysis**	81
6.1	Introduction	81
6.2	Performance parameters	81
6.3	Maintainability	82
6.4	Analysis of the parameter profile matrix	84
6.5	Application of the method: case study	85
	6.5.1 Plant items and raw data	85
	6.5.2 Data analysis	85
	6.5.3 Observations on the results	85
6.6	Discussion	87
6.7	Summary	88

xii   Contents

Chapter 7	**Failure Mode Analysis**	91
7.1	Introduction	91
7.2	Failure mode and maintenance analysis (FMMA)	91
	7.2.1 Outline method	91
	7.2.2 FMMA in concept design	92
	7.2.3 FMMA in detail design	93
	7.2.4 FMMA and condition monitoring	94
7.3	Risk and risk assessment	96
7.4	Failure mode and effect analysis (FMEA)	98
7.5	Fault tree analysis (FTA)	104
7.6	Hazard and operability (HAZOP)	106
7.7	Summary	107

Chapter 8	**Specifications, Contracts, and Management**	109
8.1	Introduction	109
8.2	Design specifications: general principles	109
8.3	Maintainability and reliability objectives	110
	8.3.1 Meaningful statements	110
	8.3.2 Relationships between design objectives	111
	8.3.3 Understanding requirements	112
8.4	Contents of a specification	113
	8.4.1 Quantitative requirements	113
	8.4.2 Maintenance and operating instructions	114
	8.4.3 Maintenance actions in design specifications	114
	8.4.4 Design review	115
	8.4.5 Requirements for the use of specific design methods	115
8.5	Changes to design specifications	116
8.6	Demonstrating maintainability and reliability	117
8.7	Standards	117
8.8	Responsibility for breakdowns in contracts	118
8.9	Management and control of design projects	118

Chapter 9	**Concept Design**	121
9.1	Introduction	121
9.2	General principles	122
	9.2.1 Elements of concept design	122
	9.2.2 Ideas	123
	9.2.3 Description of ideas	124
	9.2.4 Demonstration of feasibility	125
	9.2.5 Evaluation and choice	126

9.3	Principle of strong concepts	126
9.4	Concept development with respect to maintainability and reliability	127
	9.4.1 Stages of development	127
	9.4.2 Adaptive creativity	127
9.5	Final choice	129
9.6	Brief case study	129
	9.6.1 Problem	129
	9.6.2 Concept development	130

Chapter 10 Equipment Design Principles for Maintainability and Reliability 135

10.1	Introduction	135
10.2	Some general, qualitative guidelines	135
10.3	Load and strength	138
10.4	Reliability critical dimensions	139
10.5	Examples of design	140
	10.5.1 Fasteners	140
	10.5.2 Access to bearings for condition monitoring	142
	10.5.3 Bearing modules and magnetic drive	145
	10.5.4 Pipe joints	147
	10.5.5 Valves	147
	10.5.6 Multi-function connection	152
	10.5.7 Centrifuge	153

Chapter 11 Design for Reliability 157

11.1	Introduction	157
11.2	Component strength and applied load	158
11.3	Performance and reliability	159
11.4	Equal strength (weakest link) principle	160
11.5	Identification of the most reliable solution	161
11.6	Quantification and measurement against constraints	162
11.7	Determination of the most reliable design	164
	11.7.1 Basic method	164
	11.7.2 Comparison of designs	164
11.8	Optimization	165
11.9	Summary	167

xiv Contents

Chapter 12 Design Actions to Reduce Ongoing Maintenance Costs — 172
- 12.1 Introduction — 172
- 12.2 Surveys — 172
  - 12.2.1 Operating records — 172
  - 12.2.2 Spares usage — 172
  - 12.2.3 Maintenance personnel job records — 172
  - 12.2.4 Personnel interviews — 172
- 12.3 Decision making, allocation of resources — 173
  - 12.3.1 Objective — 173
  - 12.3.2 Costs — 173
  - 12.3.3 Design actions and outcomes — 174
  - 12.3.4 Resource allocation — 174
- 12.4 Creative problem solving — 175
- 12.5 Some problems are not designers' problems — 177

Chapter 13 The Feedback of Information to Design — 179
- 13.1 Introduction — 179
- 13.2 The use of plant data — 180
- 13.3 Data collection — 181
- 13.4 A data feedback system — 183
  - 13.4.1 Principles — 183
  - 13.4.2 The design reference — 183
  - 13.4.3 Output reports — 185
  - 13.4.4 Database information — 186
  - 13.4.5 Compiling the Design Reference — 188
  - 13.4.6 Discussion — 188

Appendix 1 Condition Monitoring — 191
- A.1.1 Introduction — 191
- A.1.2 Vibration monitoring — 191
- A.1.3 Acoustic emission — 192
- A.1.4 Displacement transducers — 193
- A.1.5 Temperature measurement — 193
- A.1.6 Lubricant monitoring — 194
- A.1.7 Corrosion monitoring — 194
- A.1.8 Electrical parameters — 194
- A.1.9 Manufacturing and process parameters — 194
- A.1.10 Crack detection — 195

Appendix 2 Creativity and Creative Problem Solving — 197
- A.2.1 Introduction — 197
- A.2.2 Creativity — 197
  - A.2.2.1 Definitions — 197
  - A.2.2.2 The designer — 198
  - A.2.2.3 The creative process — 200
  - A.2.2.4 The product — 201
  - A.2.2.5 The environment — 201
- A.2.3 Creativity tools — 202
  - A.2.3.1 An overview — 202
  - A.2.3.2 Brainstorming — 203
  - A.2.3.3 Morphological analysis — 204
  - A.2.3.4 Brainwriting — 204
  - A.2.3.5 Invitational stems, wishful thinking — 205
- A.2.4 Discussion — 206
  - A.2.4.1 Creativity — 206
  - A.2.4.2 Creative style — 206
  - A.2.4.3 Invention and innovation — 207
  - A.2.4.4 Designer(s), process, product, and environment — 207

Appendix 3 Mean Failure Rate Data — 211

Index — 213

Acknowledgements

I should like to acknowledge the contributions of my many colleagues in industry and academe who have helped shape my thoughts on this subject. To name individuals would be unfair on those omitted. Thanks are also due to the students who have undertaken projects with me in this field.

Much of the underpinning research for this book has been carried out with industry and their support is gratefully appreciated. The Engineering and Physical Sciences Research Council and its predecessors have also provided valuable resources to enable research to be undertaken and their support is gratefully acknowledged.

Foreword

Maintainability and reliability are recognized as being highly significant factors in the economic success of engineering systems and products. Also, design is the stage at which the eventual characteristics of future systems and products are determined Therefore, it is important that designers should take maintainability and reliability into account during their work. However, there is much to consider at the design stage; something has to be devised that will work and the cost and time constraints must be satisfied. There are many pressures on designers and ways have to be found to help integrate maintainability and reliability considerations into design work efficiently and effectively.

There are a number of excellent specialist text books on maintainability and reliability. The problem for designers is that they are mainly written from the perspective of the dedicated maintainability or reliability engineer. The books contain in-depth analytical methods that require information that is not available at the design stage and creative design is rarely considered. Therefore, whilst the needs of those engineers concerned with the analysis and operation of existing systems are well catered for by these texts, they are of limited use to designers.

This book is written entirely from a design perspective. It is for designers and for those who work in design-related functions. All activities from the derivation of design requirements, through concept design, to detail design are included. Relevant aspects of design management are also covered. The intention has been to show how to improve the performance of systems and products with respect to maintainability and reliability through design.

The design methods and procedures given are presented with respect to principles and applications. Some, or all, of the methods can be incorporated or a complete design procedure may be taken up. Therefore, the text can be used both as an overall study of how to achieve good maintainability and reliability through design or as a reference text for particular design methods.

Dr Graham Thompson
UMIST
1999

Chapter 1

Introduction

Maintainability and reliability are important design variables. By their decisions and actions, engineering designers largely determine the maintainability and reliability characteristics of engineering systems and products. Although good maintenance management and engineering can improve a problematic situation, it is through good design that systems and products can be created that have inherently good maintainability and reliability characteristics. The design activities that influence maintainability and reliability range from the definition of initial requirements in a specification, through concept design, to detail design. They involve design at different levels from system considerations to component selection. There are also factors to consider such as the feedback of plant experience to design, design review methods, maintenance engineering technologies, risk and safety and the effective use of design contractors. The achievement of good reliability and maintainability through design involves careful thought and action across a wide range of topics.

The objective of this book is to approach the subject of design for maintainability and reliability from a clear, distinct design perspective. At all times, the view taken will be that of an engineering designer.

The problems faced by designers are varied and difficult. Something has to be designed that works and fulfils all its required duties. Once designed, the fruits of the designers' labours are there for all to criticize. Anyone can criticize, say this or that should have been considered or point out bad features, but designers carry the burden of responsibility to create in the first instance. Therefore, given the multiplicity of problems facing designers, it is important that any methods proposed to consider maintainability and reliability in design should integrate readily with normal design activities. The approach taken in this book will not be an idealistic one in which maintainability and reliability are singled out to the exclusion of other significant design variables. To do so would be wrong and would immediately invite criticism as being unrealistic and impractical. Of course, this book concentrates on design for maintainability and reliability leaving other texts to deal with stress analysis, manufacturing methods etc. However, the methods and procedures that are presented here are compatible with other design considerations.

Design practices vary considerably from company to company. Some organizations have structured design processes that have been developed over many years. Others take a more flexible approach and adopt particular procedures and methods readily as the situation demands. Large companies face different problems to smaller companies. An inflexible, prescriptive approach to design for maintainability and reliability would find few enthusiasts and be difficult to put into practice in many situations.

The design methods and procedures given in the following chapters are presented with respect to principles and applications. Some, or all, of the methods can be incorporated within most existing company practices. It may be that specific methods are adopted, e.g. an evaluation method for choosing equipment, or that general principles are applied during design activities, e.g. the application of creativity principles during concept generation studies. Possibly, a complete design procedure may be taken up, e.g. the comprehensive design review procedure. Therefore, the text can be used both as an overall study of how to achieve good maintainability and reliability through design or as a reference text for particular design methods. It should be possible for most companies to adopt most of the design methods presented; no company design structure should be so rigid as to exclude good design practice.

Maintenance costs, caused by poor reliability and/or maintainability, are evident in many industries and applications. In some cases, the maintenance costs incurred over the life of a product or engineering system may well exceed the initial cost of manufacture. This has led to a study of life cycle costs as a way of achieving an overall economically advantageous project. In the 1970s and 1980s 'Terotechnology' emerged as a unifying approach. (Terotechnology is defined as: a combination of engineering, management, financial and other practices applied to physical assets in the pursuit of economic life cycle costs.) Terotechnology would integrate considerations of finance, manufacture, design, maintenance engineering, production operations and decommissioning. A completely integrated approach proved elusive and, although the use of the term fell away, research into the component parts of Terotechnology continued in earnest. The principle of economic life cycle costs remains as sound as ever in the 1990s and present day interest in the achievement of economic life cycle costs is strong. The objective of economic life cycle costs will no doubt continue to be the underlying theme of engineering projects in the year 2000 and beyond.

Life cycle cost models have been proposed that include capital, maintenance, downtime and decommissioning costs, but generally their practical use in design is limited. In a long life project, e.g. a process plant design, there is much uncertainty associated with the information available at the design stage. Projections for product demand, and hence downtime costs, in a process plant may prove inaccurate in a few years as the world industrial climate changes. Or, in the case of the design of an item of equipment such as a pump, the item could be used in a variety of applications in different companies and processes. The actual maintenance costs generated would depend greatly on the end users, their maintenance procedures and the care of the product by operators and not fall under the direct control of the designer. However, the difficulty in using life cycle cost models does not absolve the designer from consideration of life cycle costs.

The serious treatment of maintainability and reliability in design studies makes a significant positive contribution to the achievement of economic life cycle costs. The philosophy of this text is to focus attention on matters that are within the designer's control, to consider the many factors that directly and significantly influence maintenance costs and to propose practical methods that can be integrated into design studies. In this way, the designer achieves the best results that are reasonably practicable.

Design for maintainability is concerned with achieving good designs that consider the general care and maintenance of equipment and the repair actions that follow a failure. Design for reliability is concerned with achieving good designs that will perform a specified duty without failure.

The scope of the subject matter presented in the book is broad and the principles apply to engineering design practice in a wide variety of industries. Opportunities to influence maintainability and reliability in design practice are identified and specific design methods are given. The examples used to illustrate the principles are taken from different industries and include equipment design, manufacturing plant design and aspects of professional design practice, e.g. contractual agreements. The achievement of good maintainability and reliability depends on high quality professional design practice covering a wide range of design activities. Maintainability is not something that can be left to detail designers to check 'if the spanner will fit'. Reliability should not be left as an exercise for the analytical specialist after the detail design work has been completed because, unless the design is acceptable, or hopefully can be made so by adjustments to certain details, an expensive redesign will be required.

Design is a challenging but enjoyable subject. There is nothing more satisfying in professional practice than seeing one's ideas working. Consideration of maintainability and reliability need not detract from the enjoyment of design. Indeed, it can add to the challenge if it is approached in a positive, creative manner. Therefore maintainability and reliability need not be perceived as yet another burden for the designer. Instead, the challenge is to integrate these important variables into everyday design activities in order to produce better designs.

Chapter 2

Opportunities to influence design

2.1 Engineering Design

It is necessary to understand what is involved in engineering design activities and the nature of design work in order to identify opportunities to influence design so that maintainability and reliability may be properly considered. The characteristics of design activity are quite different from many engineering practices. Design is an open-ended activity – there is usually more than one feasible solution and different solutions may be preferred by different engineers. The combination of qualitative, quantitative and subjective factors make design both interesting and difficult. In the course of a design project a design could expect to have to deal with the following:

- uncertainty, decision making
- value judgements
- optimization (intuitive or mathematical)
- communication
- interdisciplinary activity
- engineering analysis
- inventiveness, creativity

The designer needs to be able to handle a variety of variables at one time and to have a broad engineering knowledge. It is essential to see all aspects of a problem and to make balanced judgements to prioritize conflicting requirements. Divergent thinking skills are important in order to generate possible solutions. Equally important are convergent thinking skills to evaluate and focus down to a particular solution.

2.2 Design activity

2.2.1 *Design phases*

There are three major phases of design activity that can be clearly identified in a design project, large or small:

- definition of requirements in a specification
- conceptual design
- detail design

Detail design covers a wide range of activities from schematic layouts to the preparation of manufacturing drawings. It is possible to describe design activity in great depth using complex design models of design in which the above phases are subdivided into discrete activities. For example, the concept design stage comprises an ideas generation activity and an evaluation activity.

Each phase of design presents particular opportunities to ensure that maintainability and reliability are considered properly. Different approaches are required at each phase since different design skills and activities are used. Each phase and its constituent activities will be considered in depth and appropriate design methods and procedures described. An overview is given below to give an overall perspective of the scope of the work carried out in a design project.

2.2.2 Definition of requirements in a specification

The first stage is to understand the objectives of the design project and to prepare a design specification that clearly describes and defines the design requirements. This is particularly important when design work is undertaken under contract where specific parts of a total specification may be used for contracted out design work. Specifications vary in type; the starting point is the requirements of the client and a set of engineering functional requirements are derived from a client's requirements. These should be explored by the designer to ensure that all is understood. Care should be taken not to include solutions into the specification since that may over-constrain design activity. Also sufficient flexibility needs to be incorporated to permit the search for a good solution. Although design requirements are defined at the beginning of a project, they are not necessarily fixed permanently. With the agreement of the client, the design specification can be altered. After all, the objective is to produce the best job for the client which may mean making changes as better options are identified during design.

Design specifications contain both qualitative and quantitative statements and present an opportunity to define maintainability and reliability requirements right at the outset of an exercise. The specification is used as a reference point throughout a project, particularly in design review activities, and at the handover stage at the completion of a large project.

Therefore, the derivation of a good definition of maintainability and reliability requirements in a design specification is very important.

2.2.3 Concept design

The concept stage is that in which ideas are generated and evaluated to best satisfy the specification. At this stage the designer needs to use creative thinking to generate ideas and, separately, judgemental thinking to evaluate them. Many design projects lack success because not enough ideas were generated. It demands strong will-power to keep thinking broadly when an apparently good idea has been suggested. Too often, designers become focused on a single idea early in a design project. Evaluation should be objective. A danger is that an evaluation exercise is undertaken that justifies the preferences of the designer. A good evaluation exercise should assess ideas and concepts objectively with respect to the requirements defined in the specification.

It is possible in concept design to generate and develop concepts that are inherently good in certain respects. For example, a bridge structure that is conceived from carbon fibre reinforced plastic will lead to a stiff, light structure although it may well have other drawbacks such as cost. It is possible to generate concepts that are inherently unreliable; they may embody technology working close to or just beyond normal working limits or have a multiplicity of parts. Conversely, it is possible to generate concepts that contain features that will lead to good maintainability and reliability and the generation of such 'strong concepts' is a sign of good design.

2.2.4 Detail design

In detail design, the chosen design concept is developed in some depth and takes shape as general arrangement drawings. Once ideas have been developed to this stage, more detailed analyses (stress, fluids etc.) can be performed to optimize the design and refine the ideas further. In a large project, it may be necessary to decide early and develop only one concept in detail. In smaller projects, two or more concepts may be developed and one chosen later for final detail design as the best design emerges.

The final design and drawing work is undertaken to produce drawings from which manufacture can take place. It is interesting to observe that some complex engineering analyses can only be performed when the fine detail is known, thus, their contribution during the design process is minimal. Their significant contribution is a final check that some component or system will perform its required duty.

The quality of detail design can make or break a good concept. Attention to detail is vital if reliable equipment is to be produced. For example, poor component quality leads to low reliability, badly located parts may fail, threads left exposed to corrosive environments may well fail or be difficult to remove during maintenance. The selection and/or location of small fasteners may cause maintenance problems. Professional engineering designers have the responsibility to direct detail designers in their work.

2.3 Levels of design

It is necessary to clarify the difference between levels of design and terms such as 'concept' and 'detail'. Levels of design can be conveniently described as:

system
  sub-system
    equipment design or functional unit
      sub-assembly
        component

In the case of large systems, it may be necessary to divide the system into sub-systems before moving down to the levels of equipment design and component design. Similarly, large machines may have definable sub-assemblies. More than four levels may become unmanageable and often three levels of design are perfectly adequate because, in the case of smaller systems, sub-systems and sub-assemblies may not be necessary. When deciding on the levels to describe a system, it is sometimes convenient to take the lowest level of detail as that of the maintainable or replaceable item.

Therefore, the number of levels should be considered on a case-by-case basis so that the system can be adequately described without over-complication. It should be recognized that scale plays an important part because a machine in a large system may well be defined as a functional unit whereas a similar device may be considered to be a small system in another situation. For example, a machine to degrease components in an assembly plant would be defined as a functional unit but a power jet car washing device may well be defined as a small system in product design. Also, in the definition of levels in one system, anomalies may arise. For example, one functional unit may divide down immediately to the component level while another may require sub-assemblies. There is no universal categorization, but rather an appeal to common sense is required and uniformity should not be expected.

System design does not equate to concept design, nor does functional unit and component design equate to detail design. In the design of a system, a definition of system requirements are required in a specification. Alternative system concepts are then considered and, following detail design, the system, its inputs/outputs and all the functional relationships and requirements within the system to convert inputs to outputs are defined. In the design of functional units, the system design specifies the functional requirements, alternative concepts are then considered and the equipment then designed in detail.

Consideration of the levels of design identifies opportunities to influence design activities with respect to maintainability and reliability but it is essential to take account of design level carefully. Designers working at the system level have different priorities to designers working at the detail level. If one bombards the system designer with requests to consider maintainability and reliability in terms that are appropriate to component level then those appeals will be ignored. Design methods to consider maintainability and reliability must be devised so that they are appropriate to particular design levels. It will be seen that in Chapter 4, Section 4.5 System Review, specific actions can be taken at the systems level that focus attention on equipment level design in certain key areas. Special functional unit analyses can be used to determine the usefulness of a functional unit to the system of which it is a part, and the usefulness includes maintainability and reliability. Particular sets of components can be identified for consideration in a design review.

2.4 Design models

2.4.1 Equipment and product design

In equipment design, the three phases of Specification, Concept and Detail design can be illustrated by the simplified model shown in Fig. 2.1. Specific activities are identified in the three phases and each activity represents an opportunity to consider maintainability and reliability.

2.4.2 Large scale projects

Large scale projects are too complex to be modelled as simply as shown in Fig. 2.1. The actual process followed will vary from project to project, but the main elements will be as shown in Fig. 2.2.

The client's requirements are defined in a specification and then there usually is a tender stage. The tender stage involves further investigation of the requirements during which time certain factors may well emerge that influence maintainability and reliability. For example, particularly stringent maintainability requirements for hazardous materials or an arduous working environment that affects reliability may be found. The

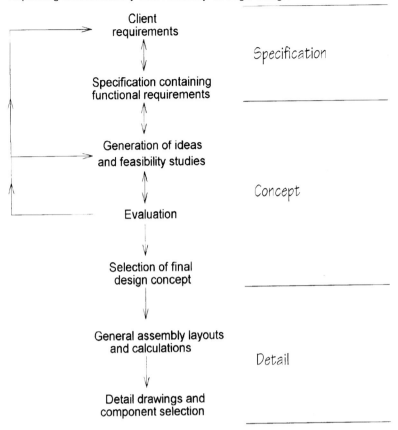

Fig. 2.1 Design model: equipment

tender should reflect such matters rather than leaving them until later when time and money are short. Once the contract has been placed, the system design will be undertaken which leads to the definition of sub-systems and function units that comprise the system. The system design activity involves concept design which involves the generation and evaluation of ideas. Once functional units have been defined, then the equipment is designed or selected to fulfil the functional requirements. Individual functional unit design activity can be modelled as shown in Fig. 2.1.

Therefore, in a large project, there are many concept design activities at the system and functional unit levels. There are also many detail design phases. All these activities present many opportunities to make sure that maintainability and reliability are properly considered (or opportunities to neglect them).

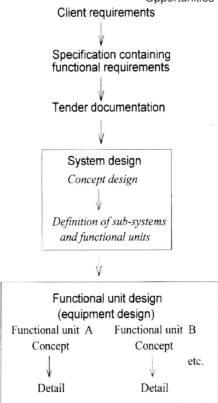

Fig. 2.2 Design model: systems and equipment

2.4.3 *Iteration in design*

It is worth reflecting on the iterative nature of design work. Figure 2.1 indicates the possibility of iteration between activities in the early stages of design but not from the latter stages back to concept design. It is worth considering briefly the possibility of iteration back to the concept stage from detail design because it will be seen that the importance of good concept design is highlighted. In the case of small scale projects, it is possible that one could iterate back from a detail design activity to a higher level of detail design, perhaps to alter the general arrangement of parts. It is even conceivable that one might even iterate back to alter the concept if detail design proves really unattractive.

However, in the case of large scale projects the case is quite different. Once the concept stage has been concluded and a decision made to proceed further with a particular concept, then very large resources are committed. If problems occur in the latter stages of detail design then it is feasible to iterate back to revise general arrangement drawings, say to devise another scheme to fulfil a set of functional requirements. However, it is highly unlikely that an iteration back to the concept stage is possible because to do so would be very damaging economically. Therefore, iterations back to the concept are only possible for small projects, unless extremely serious problems are encountered that would cause the project to fail.

Consequently, the importance of high quality concept generation and evaluation cannot be overstated.

2.5 The nature of design activity

2.5.1 Divergent and convergent thinking

Divergent thinking is used in many design activities. Examples include: to generate ideas; to search for information; to think up different combinations; and to redefine problems. The list is not exhaustive. Divergent thinking is a difficult discipline to master. For success, the designer must suspend judgemental thinking to leave the mind free and unfettered to explore possibilities and to let the imagination generate ideas.

Convergent thinking involves judgemental thinking. It involves: evaluation; decision making; making sense of information; ranking priorities; and making value judgements. There are pitfalls in convergent thinking. The designer must be careful not to superimpose personal prejudice on objective thinking.

2.5.2 Divergent–convergent coupled activities

All design phases, from specification through concept design to detail design, involve divergent–convergent thinking. The exploration of the specification is divergent whilst the capturing of salient functional specifications in written form is convergent. The generation of ideas in concept design is divergent whilst the evaluation of concepts and the making of choices is convergent. At the detail design phase, divergent thinking is involved in component synthesis, consideration of materials etc. whilst convergent thinking is the decision making in the choice of materials etc.

One might even consider the whole of the design process to be divergent–convergent: starting from the client's initial requirements through concept design involving consideration of alternatives to the choice of a final detail design. Therefore, design activity involves divergent–convergent thinking at all levels of design and within an overall divergent–convergent process as illustrated in Fig. 2.3.

Recognition of the existence of divergent–convergent thinking is significant in identifying opportunities to influence reliability and maintainability. At different stages and at different levels of design, the objective is to devise practical ways to influence designers' divergent

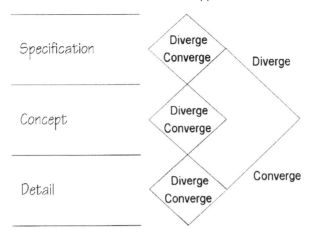

Fig. 2.3 Divergent–convergent thinking in design activities

thinking to include maintainability and reliability. Similarly, ways to include maintainability and reliability in convergent thinking need to be found. In the cases of divergent and convergent thinking, the design methods required to influence design should be appropriate to the stage of design considered.

2.6 Feedback of experience to design

There are also opportunities to improve the maintainability and reliability of products and manufacturing plant after use. Feedback from customers will identify shortcomings in products and particular design improvements can be made accordingly. The design changes required will be product-specific and focus on particular failure mechanisms (e.g. excessive wear or fatigue), accessibility problems, frequency of adjustment etc.

In the case of manufacturing plant, there will be feedback from production departments that identify sources of poor maintainability and reliability. A systematic approach could be adopted to actively collect and analyse data. Collection and analysis of data provides valuable information for the design of future plants. If it is required to make improvements to existing production plant, then it is likely that there will be insufficient design and/or financial resources available to deal with all cases in an optimum way. Therefore a carefully considered decision making approach is required to effect improvements to achieve the best overall results with the resources available.

In some cases it will be recognized that, in general, a plant has poor maintainability and/or reliability but specific details may not be recorded.

A plant review can then be used to identify the causes of poor performance in this respect and specific actions taken to remedy the situation. The causes may be instances of poor equipment performance or more generic problems that affect the whole plant. The former is more easily dealt with, but plant reviews should identify all types of problem.

2.7 Summary

Design activities may be characterized by their multifaceted nature, uncertainty, and combinations of qualitative and quantitative factors. Design activities are varied and encompass the definition of requirements in a specification, concept design and detail design. Also, design is carried out at different levels from system design to component selection.

Therefore, there are many opportunities to influence design with respect to maintainability and reliability. Accordingly, many different methods are needed that are appropriate to each design activity. There cannot be a single approach to design for reliability that is useful in all cases. A portfolio of methods is required so that the correct approach can be used at each stage.

Design methods must also reflect the type of thinking used by designers. If divergent thinking is involved, then a design method involving judgemental thinking would be completely inappropriate. Care must be taken not to interfere and spoil the design work being undertaken. In a convergent design activity, sensible evaluation and decision making to include maintainability and reliability is called for so that design activities are not diverted off-track.

Following the design, manufacture and use/operation of products and systems, there are still opportunities to effect improvements with respect to maintainability and reliability. Feedback from customers and production departments identifies specific areas where attention can be focused. In cases where there is a recognition of general poor performance with respect to reliability and maintainability, then design review methods can be used to identify the best ways to make improvements.

Therefore, design opportunities to achieve good maintainability and reliability extend from the definition of requirements in the initial specification, through all phases and activities of design, to actions that can be taken during the operating life of products and systems. Different methods by which maintainability and reliability may be considered in design activities are presented in the remaining chapters. Each has an appropriate place in design, while some may be used in different situations. Good maintainability and reliability can be achieved through design and success is achieved by using appropriate design methods.

Bibliography

The following texts provide a general discussion of design activities.

Pugh, S. (1991) *Total design* (Addison-Wesley).

Jones, C.J. (1970) *Design methods* (John Wiley).

Cross, N. (1994) *Engineering design methods* (John Wiley).

Pahl, G. and **Beitz, I.** (1996) *Engineering design a systematic approach* (Springer).

Siddall, J.N. (1972) *Analytical decision making in engineering design* (Prentice-Hall).

Chapter 3

Maintainability and Reliability: Basic Principles

3.1 Introduction

Reliability theory has been developed to a great depth. It can involve complex mathematics and sophisticated computational techniques and much of the work is concerned with the modelling of complex systems. Reliability engineering also has been the subject of much study, topics ranging from understanding the causes of failure in components (e.g. crack propagation, wear) to understanding the failure characteristics of aircraft, large manufacturing systems etc. The results of reliability engineering studies lead to decisions about maintenance practices and hence there is much interaction between reliability engineering and maintenance engineering (e.g. Reliability Centred Maintenance). Maintenance engineering and management has itself been the subject of in-depth study. Further applications of reliability lead on to safety assessment and the subjects of reliability, maintenance and safety are intertwined.

In-depth treatises on reliability and maintenance theory are inappropriate for engineering design, they are directed primarily at the reliability and maintenance specialists. A lot of work deals with the analysis of existing systems whereas the designer is concerned with the creation of the system in the first place. However, just as in most engineering analysis subjects, a designer requires knowledge of the basic principles in order to be effective. Therefore, an introduction to the basic principles that are particularly relevant to design is given here. For further study of reliability and maintenance principles, the reader is referred to the texts given at the end of this chapter.

3.2 Definitions

Reliability is the probability that a component, device or system will continue to perform a specified duty under prescribed environmental conditions for a given time. The definition applies equally to components, small and large machines, and manufacturing systems. Environmental conditions include such factors as temperature, vibration, dirt etc. Operating circumstances are significant, for it is well known that certain machinery will prove more reliable when operated by personnel who understand the equipment and are sensitive to its use. Sophisticated equipment designed for use by skilled persons will be unreliable in untrained hands. Poorly maintained equipment will also be unreliable. The above definition of reliability assumes that the equipment will be operated and maintained correctly.

Reliability is a function of time. Designers are well used to designing for a given life, therefore the concept does not immediately present any problem. However, too often one hears references to a reliability of x percent without reference to time. In the definition of reliability requirements, the expected design life is a significant factor.

The definition of reliability refers to a specified duty that is required to be performed. This may be defined in various ways which would affect design and there may even be different requirements for the same device. For example, a self-inflating life-raft could have a reliability defined with respect to the probability of it deploying correctly when required, say after being left for six months. The same life raft could also have a probability of remaining intact in a force 9 gale for 3 weeks, in temperatures down to $-2°C$ etc. A manufacturing system may have a reliability defined with respect to its production capacity, e.g. output not falling to less than 50 percent for two days in a 12 month period. Alternatively, production quality may be the issue and reliability may be defined with respect to the purity of a product. Within a system, a machine may have a reliability requirement to control, say, the temperature of a liquid between two limits or perhaps to supply steam at a certain temperature, pressure and flow rate.

This brief discussion of the specified duty emphasizes the need to define *failure*. Failure is the condition when the performance of a component, device or system becomes unacceptable. Often, there is no single criterion that can be used alone to define failure except in the case of a very simple component. Take the case of a railway locomotive and carriages. A failure criterion may be defined with respect to its ability to complete a specified number of journeys of a certain length. However, a very different view would be taken of its failure criterion with respect to a catastrophic event that could lead to loss of life or injury. Even for a small device there can be different definitions of failure that influence the designer's choice of components. In the case of a bolted flange, one failure criterion may refer to a requirement for a minimum continuous leak rate whereas another could refer to a single large leakage event. The first criterion affects the choice of gasket whereas the second influences the choice of flange and bolt materials.

Failure need not refer to safety. It refers to the ability of a component, device or system to perform a specified duty. The subject of failure criteria and optimum design for reliability is discussed in Chapter 11.

With respect to *maintainability*, the designer has to take a different view from that of the maintenance manager. One can define maintainability with respect to the probability that a device or system can be returned to a specified condition using pre-specified practices within a specified time. Such a definition leads to mathematical analyses using repair rates in a similar manner to reliability analyses. Although useful to the maintenance manager in the analysis of data accumulated in service, this approach is not very useful in design. The designer should consider those factors which are under his/her control. The time taken to return a failed machine to a working condition includes fault finding, carrying out the repair, organizing manpower and obtaining spares. The time taken to identify the fault and repair it, that is time worked on the job using tools, is usually less than half that of the total repair time and commonly much less than that. The design of the equipment influences fault finding, dismantling and reassembly and adjustments.

Therefore, it is useful to differentiate between:

- *mean time to repair*, which is the total time required to return a machine to a satisfactory working condition, and
- *mean corrective repair time*, which is the time taken to identify the fault, carry out the repair and adjustments assuming that all tools, spares and required manpower are available.

The designer should be primarily concerned with the mean corrective repair time.

3.3 Maintainability prediction in design

Consider the case of a machine for which there is no history of performance available. It may be a machine that has just been designed or it may be a proprietary device that is being considered for selection as part of a larger machine or system.

The failure modes of the machine are first identified. In the case of a small machine, these will be the individual components of the machine.

For each failure mode, a corrective repair time is estimated to bring the machine back to a satisfactory working condition. The time should include component removal and replacement including adjustments. The repair may be done by replacing a defective component with a new one, or by removal and carrying out a bench repair before replacement. In the former case it has to be assumed that a replacement is immediately available, as discussed under Section 3.2 above. But in the case when a bench repair is expected, the corrective repair time should include the time on the bench as well as the removal and replacement times from/into the machine, e.g. a valve removed to a workshop to refurbish its seat and replace seals. In some cases, failure may not mean that a component needs replacement at all, but rather that adjustments or other actions are required to bring the machine back to an acceptable working condition. For example the adjustment of the diamond cutter and tracking in a glass production process.

A set of general assembly drawings is usually sufficient to estimate repair times, but care is needed to consider the environment in which the machine will be used. External corrosion of fasteners, the location of a machine within a plant, noise levels, heat etc. will all affect the abilities of maintenance personnel. Advice from experienced personnel is key to the determination of meaningful results. Unfounded guesses should never be used.

Let each failure mode have a corrective repair time m_i (which includes the repair times of any secondary failures). When a failure occurs, the corrective repair time M of the machine will be one of the values m_i. Assume that the probability of failure of each failure mode is F_i. If a failure occurs, the probability that the ith mode will fail is: $\dfrac{F_i}{\Sigma F_i}$ assuming that all the modes of failure are independent. Therefore, the expected value of the corrective repair time M is

$$M = m_1 \frac{F_1}{\Sigma F_i} + m_2 \frac{F_2}{\Sigma F_i} + \cdots m_n \frac{F_n}{\Sigma F_i}$$

where $n =$ total number of failure modes. Hence,

$$M = \frac{\Sigma F_i m_i}{\Sigma F_i} \tag{3.1}$$

Equation (3.1) gives an estimate of the mean corrective repair time of a machine, under prescribed conditions, over the life of the equipment.

It can be seen from equation (3.1) that the actual value of the probability of failure F_i is not required to perform the calculation, only the relative values of F_i are needed. Failure rate data (see Section 3.7 below) may be used to give an estimate of the relative values of the failure probabilities. Let λ_i = the failure rate of failure mode i. Then the mean corrective repair time M is given by

$$M = \frac{\Sigma \lambda_i m_i}{\Sigma \lambda_i} \qquad (3.2)$$

If failure rate data are not available, but designers have reasonable experience of machine failures, then estimates of the relative frequencies of expected component failure can be used in equation (3.2) in place of λ_i.

It is interesting to note that mean corrective maintenance time and reliability are not independent in machine design. The probability of failure of components determines the failure mode and hence the required corrective action.

3.4 Availability

Simple steady state availability A is defined as

$$A = \frac{\text{the time a machine or system is (or is capable of being) in operation}}{\text{total time the machine or system is required to be in operation}}$$

$$A = \frac{\text{total 'up time'}}{\text{total 'up time'} + \text{total 'down time'}}$$

$$= \frac{\text{mean time to failure}}{\text{mean time to failure} + \text{mean time to repair}}$$

The above definition assumes that there are two states: working and failed.

Note that the steady state availability calculation gives the same result for machines that have different failure characteristics. For example a machine that, on average, runs continuously for four days and then requires a one day cleaning/refurbishment stoppage has an availability of 80 percent. The same result is obtained for a machine that stops on average for a five minute period after a run of 20 minutes. Clearly these two machines have different characteristics concerning the frequency of failure and repair time that the availability calculation does not reveal.

Production systems can have intermediate states which can be incorporated into an availability calculation. The production availability A_{prod} is

$$A_{prod} = \Sigma_{all\ i} \frac{(availability)_i \times (output\ level)_i}{maximum\ possible\ output\ level}$$

For example, assume that a production system has a maximum possible output of 500 units per shift but actual production output is recorded as 400 units per shift with an availability of 70 percent and 200 units per shift with an availability of 25 percent. There is no output for 5 percent of the time. The production availability is then

$$A_{prod} = \frac{400 \times 0.7 + 200 \times 0.25 + 0 \times 0.05}{500} = 0.66$$

$$A_{prod} = 66\%$$

3.5 Reliability and failure rate

3.5.1 Reliability

Assume a population of N_0 nominally identical components have been tested. After time t the number surviving is N_s and the number of failed components is N_f.

$$N_o = N_s + N_f$$

The probability of a component surviving to time t is $\frac{N_s}{N_o} = R(t)$ where $R(t)$ = the reliability of the component which is a function of time. The probability of failure is $F(t) = \frac{N_f}{N_o}$

$$R(t) + F(t) = 1$$

The hazard rate is defined as

$$h(t) = \frac{dN_f}{dt} \cdot \frac{1}{N_s} \tag{3.3}$$

The hazard rate $h(t)$ is sometimes called the instantaneous failure rate or mean failure rate (sometimes other terminology is used). The total numbers failing per unit time $\frac{dN_f}{dt}$ is not itself a useful quantity since such a failure rate would depend upon the number of components in service, i.e. two tests using different numbers of components would give different 'failure rates'.

Now

$$F(t) = \frac{N_f}{N_o} = 1 - R(t)$$

therefore,

$$R(t) = 1 - \frac{N_f}{N_o}$$

differentiating with respect to time,

$$\frac{dR(t)}{dt} = -\frac{dN_f}{dt} \cdot \frac{1}{N_o}$$

and since $\dfrac{dN_f}{dt} = h(t).N_s$ we can write

$$\frac{dR(t)}{dt} = -h(t)\frac{N_s}{N_o}$$
$$= -h(t).R(t)$$

Rearranging terms gives

$$\frac{1}{R(t)}.dR(t) = -h(t).dt$$

Integrating,

$$\int_1^R \frac{dR(t)}{R(t)} = -\int_0^t h(t).dt \qquad (3.4)$$
$$R(t) = e^{-\int_0^t h(t).dt}$$

Note the integration limits. The component is assumed to be perfectly reliable at the outset therefore $R = 1$ at $t = 0$.

In order to calculate the reliability of a component, we require knowledge of the hazard rate (mean failure rate) $h(t)$ and the design life t. The term 'failure rate' is in very common usage in the engineering community and is more often used than hazard rate, instantaneous failure rate or any other precise terminology. However, what is meant when people discuss failure rates is the mean failure rate as defined by equation (3.3), which is a meaningful quantity, and not simply $\dfrac{dN_f}{dt}$.

3.5.2 *Mean failure rate*

Many components have a variable mean failure rate throughout their life. Generally, if a set of nominally identical components are tested, they exhibit a high mean failure rate initially, then the mean failure rate falls to

24 Improving maintainability and reliability through design

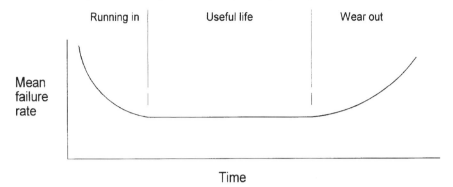

Fig. 3.1 Mean failure rate as a function of time

a significantly lower level before increasing again towards the end of the design life. The usual way to illustrate this behaviour is as shown in Fig. 3.1.

The initial phase is the 'Running In' phase. Here, the failures are caused by undetected manufacturing errors, incorrect assembly procedures and the use of materials that are not to the required specification etc.

The second phase is commonly referred to as the 'useful life' phase. Here the mean failure rate is low and is constant for the majority of the life of component. Failure is due to chance coincidences of high loads and low component strength combinations.

The last phase is the 'Wear Out' phase in which the mean failure rate increases as components reach the end of their design life and fail due to numerous mechanisms, e.g. fatigue, wear, creep, contamination, corrosion.

The characteristic given in Fig. 3.1 is a broad generalization which applies more often to electrical components than to mechanical ones. In the case of mechanical components, the extended period of low, constant mean failure rate is not evident to the same degree although the mean failure rate does fall during a Running In period and rises in the Wear Out phase. Also, studies of particular components in service may reveal no Running In phase if pre-service testing and strict quality checks have been made during manufacture and assembly. However, in reliability prediction, an assumption of constant mean failure rate is commonly made for many components.

3.5.3 *Constant mean failure rate*
If the mean failure rate is assumed constant, then

$$h(t) = \text{const} = \lambda$$

Substituting in equation (3.4)

$$R(t) = e^{\int_0^t -\lambda.dt}$$

$$R(t) = e^{-\lambda.t} \tag{3.5}$$

Equation (3.5) is one of the most commonly used equations in reliability prediction. If the mean failure rate λ can be estimated then the reliability can be predicted for a given design life t. In many cases an assumption is made that the mean failure rate is assumed constant for both mechanical and electrical components. While it is recognized that this is an approximation for mechanical components, nonetheless many reliability and analysis/prediction procedures are carried out on this assumption.

3.5.4 Mean time to failure, MTTF

$F(t)$ is the probability of failure up to time t (not at time t but in the period up to t). $F(t)$ is the cumulative function. The failure probability density function is $f(t) = \dfrac{dF(t)}{dt}$.

The probability of failure between $t = t_1$ and $t = t_2$ is $\int_{t_1}^{t_2} f(t).dt$.

The mean time to failure MTTF is the mean value of the probability density function. Therefore

$$\text{MTTF} = \int_0^\infty t.f(t).dt$$

$$\text{MTTF} = \int_0^\infty t\left(-\frac{dR(t)}{dt}\right) dt$$

and integrating by parts

$$\text{MTTF} = [t.(-R(t))]_0^\infty - \int_0^\infty (-R(t)).1.dt$$

giving the useful equation

$$\text{MTTF} = \int_0^\infty R(t).dt \tag{3.6}$$

If the mean failure rate λ is constant, then

$$\text{MTTF} = \int_0^\infty e^{-\lambda.t}.dt$$

$$\text{MTTF} = \frac{1}{\lambda} \tag{3.7}$$

Equation (3.6) may be used generally to calculate the mean time to failure, equation (3.7) applies to constant failure rate only.

Some engineers use the term 'mean time between failures' (MTBF) instead of mean time to failure.

3.6 Reliability modelling

3.6.1 Reliability modelling and design

Reliability models are used in design to highlight the critical parts of systems that are particularly important to ensure high reliability. After reviewing a reliability model it may be that certain changes are made to improve the system reliability. Chapter 4, Design Review, discusses how reliability models may be used at the systems level in a review exercise. Also, in small items of equipment, a reliability model may be made of different components to evaluate equipment reliability, see Chapter 5.

Reliability models are block diagrams that show the relationships between elements from a reliability perspective. They are based on the requirements that must be met for a system to function satisfactorily and not on the physical layout of components.

3.6.2 Series elements

A set of elements are in series reliability if all the elements must work all of the time for the system to operate. Figure 3.2 shows n elements in series.

The reliability R_s of the system is given by

$$R_s = R_1 \times R_2 \times R_3 \ldots \times R_n \tag{3.8}$$

If $\lambda =$ constant, then

$$R_s = e^{-\lambda_1.t} \times e^{-\lambda_2.t} \times e^{-\lambda_3.t} \ldots \times e^{-\lambda_n.t}$$
$$R_s = e^{-(\lambda_1 + \lambda_2 + \lambda_3 \ldots + \lambda_n).t}$$
$$R_s = e^{-\lambda_s.t} \tag{3.9}$$

where

$$\lambda_s = \lambda_1 + \lambda_2 + \lambda_3 \ldots \lambda_n \tag{3.10}$$

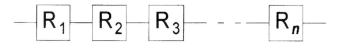

Fig. 3.2 Elements in series

The system failure rate is therefore the sum of the individual failure rates of the elements of the system. This gives rise to the *component count method* of reliability prediction which also includes maintainability prediction, see Section 3.8. The system

$$\text{MTTF} = \frac{1}{\lambda_s} \qquad (3.11)$$

3.6.3 Parallel elements: active redundancy

A set of elements are in parallel active redundancy if all the elements of the system are normally working but the system will continue to function satisfactorily provided that any single element of the system will work. Figure 3.3 shows n elements in active redundancy.

For the system to fail all the elements must fail. Therefore, the probability of failure of the system F_s is

$$F_s = F_1 \times F_2 \times F_3 \ldots \times F_n$$

But,

$$R_s = 1 - F_s$$

Therefore the system reliability R_s is given by

$$R_s = 1 - (1 - R_1) \times (1 - R_2) \times (1 - R_3) \times \ldots (1 - R_n) \qquad (3.12)$$

In order to calculate the MTTF of a parallel system, equation (3.6) must be evaluated.

For two elements in active parallel redundancy

$$\text{MTTF}_2 = \lambda_1^{-1} + \lambda_2^{-1} - (\lambda_1 + \lambda_2)^{-1}$$

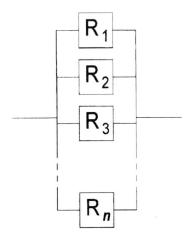

Fig. 3.3   Elements in parallel

For three elements in active parallel redundancy

$$MTTF_3 = \lambda_1^{-1} + \lambda_2^{-1} + \lambda_3^{-1} - (\lambda_1 + \lambda_2)^{-1} - (\lambda_1 + \lambda_3)^{-1}$$
$$- (\lambda_2 + \lambda_3)^{-1} + (\lambda_1 + \lambda_2 + \lambda_3)^{-1}$$

The case of *standby redundancy*, where an element remains idle until called on because of another failure, is analysed differently to full active redundancy. The situation is complicated because the standby unit may not be operable when called on, e.g. seizure of components may have occurred. For an analysis of these cases see references **(2)–(4)**.

3.6.4 Combined series and parallel

A system that comprises series and parallel elements can be reduced in complexity in stages in order to carry out a reliability analysis. As an example, Fig. 3.4(a) shows a system that comprises six elements.

The system is analysed in stages.

(i) The series elements R_1 and R_2 are reduced to a single element R_A where $R_A = R_1 \times R_2$
(ii) The parallel elements R_5 and R_6 are reduced to a single element R_B where $R_B = 1 - (1 - R_5) \times (1 - R_6)$

The system is now as shown in Fig. 3.4(b).

(iii) The parallel elements R_A and R_3 are reduced to a single element R_C where $R_C = 1 - (1 - R_A) \times (1 - R_3)$

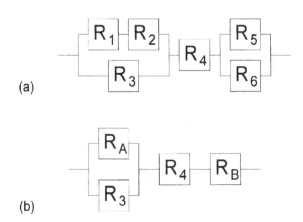

Fig. 3.4 (a) A combined series and parallel reliability model. (b) Simplified model

(iv) The system is therefore reduced to three elements in series R_C, R_4, and R_B which gives a system reliability R_s of

$$R_s = R_C \times R_4 \times R_B$$
$$R_s = (1 - (1 - R_1 \times R_2) \times (1 - R_3)) \times R_4$$
$$\times (1 - (1 - R_5)(1 - R_6))$$

3.6.5 *Partial active redundancy*

There are systems that comprise several elements which are all normally working but the system is such that it will work to a satisfactory level provided that a certain number of the elements continue to work. For example, a cooling system may comprise four pumps, all working, such that provided that any two out of the four continue to work then the system will continue to provide sufficient cooling water. Such a system is called partial active redundancy.

First take the case of a system of three components 1, 2 and 3 in which the system will work provided any two of the elements work. The possible satisfactory operating scenarios and their probabilities are shown in Table 3.1.

The system reliability R_s is therefore

$$R_s = R_1 \times R_2 \times R_3 + F_1 \times R_2 \times R_3 + R_1 \times F_2 \times R_3 + R_1 \times R_2 \times F_3 \tag{3.13}$$

If all components are identical, then

$$R_s = R^3 + 3 \times F \times R^2 \tag{3.14}$$

When the number of components is small, the system reliability can be calculated following the method shown in Table 3.1 and equations (3.13–3.14). For the general case of a system comprising n identical components which operates satisfactorily provided that no more than m of the elements may fail (the m components can be any of the n), the system reliability R_s is given by the sum of the first $m + 1$ terms of the expansion of $(R + F)^n$.

Table 3.1 Failure scenarios

	Probability
All work	$R_1 \times R_2 \times R_3$
1 fails, 2 and 3 work	$F_1 \times R_2 \times R_3$
2 fails, 1 and 3 work	$R_1 \times F_2 \times R_3$
3 fails, 1 and 2 work	$R_1 \times R_2 \times F_3$

3.7 Prediction of reliability using failure rate data

If the design life t and the mean failure rate λ are known, then equation (3.9) may be used to predict reliability. The difficulty is knowing what value of λ to use in the calculation. The best data are obtained from operating records for the same loading conditions, the same supplier of components etc., collected over a long period. Of course this ideal is rarely to hand, but data sources are available that give mean failure rate data for components that can be used in reliability prediction calculations with meaningful results.

Component mean failure rate data are available as a nominal (or base) mean failure rate λ_0 from data sources. The nominal mean failure rate is then multiplied by factors k_1 and k_2 to make the data more representative of the case under examination. The factor k_1 refers to the environment in which the component will be operated. A hostile or severe environment increases the mean failure rate significantly. The range of factors that are used for k_1 for different applications and industries is given in Table 3.2. One has to gain a 'feel' for the application since there is no hard and fast rule. For example, not all chemical plant applications are the same and judgement is needed when choosing an appropriate value of k_1. For each application a reasonable estimate has to be made so that if, say, a device is to be used in an environment which is dirty and where vibrations are present, then $k_1 = 2$ might well be appropriate. The factor k_2 refers to the 'stress level' in the component, Table 3.3 gives examples of the range of values of k_2. If a component is expected to operate well below its maximum limit, say at 40 percent of maximum, then the mean failure rate would be

Table 3.2 Environmental stress factors k_1 (Ref (**1**))

General environmental conditions	k_1
Ideal, static conditions	0.1
Vibration-free, controlled environment	0.5
Average industrial condition	1.0
Chemical plant	1.5
Offshore platforms/ships	2.0
Road transport	3.0
Rail transport	4.0
Air transport (take-off)	10.0

Table 3.3 Stress factors k_2 (Ref (1))

Component nominal rating	
%	k_2
140	4.0
120	2.0
100	1.0
80	0.6
60	0.3
40	0.2
20	0.1

Note: for equipment designed against a specific design code (for example vessels) rating = 100%; for other equipment (for example pumps, valves, etc.) rating = 80% or less may be appropriate.

reduced to 20 percent of the nominal mean failure rate (i.e. λ_0 multiplied by 0.2). Both k_1 and k_2 are applied to the nominal mean failure rate.

The failure rate can be modified further to take account of particular failure modes, if sufficiently detailed data are available. For example, if 'external leakage' was the definition of failure, then pump failures due to corrosion of impellers, shaft coupling problems etc. should be excluded. Not all data sources give such precise details of the cause of failure and in many cases the analysis proceeds without such a refinement of the data.

Therefore, the mean failure rate λ that is to be used in reliability calculations is obtained by modifying the nominal mean failure rate λ_0 as shown in equation (3.15).

$$\lambda = \lambda_0 \times k_1 \times k_2 \qquad (3.15)$$

Sources of data are given in Table 3.4. In a reliability exercise it is best if data can be used from a source that is as close as possible to the case under examination because that reduces the need for the use of an environment factor in equation (3.15).

3.8 Component count method for predicting reliability and maintainability

The component count method is a systematic method of performing calculations to assess the reliability and maintainability of machines and systems. It uses the reliability prediction equations (3.10) and (3.11) in a

Table 3.4 Sources of failure rate data (from Ref (1))

	Data source	Title	Publisher and date
1.	CCPS	Guidelines for process equipment reliability	American Institute of Chemical Engineers, 1990
2.	Davenport and Warwick	A further survey of pressure vessels in the UK 1983–1988	AEA Technology—Safety and Reliability Directorate, 1991
3.	DEFSTAN 0041, Part 3	MOD practices and procedures for reliability and maintainability, Part 3, Reliability prediction	Ministry of Defence, 1983
4.	R. F. de la Mare	Pipeline reliability; report 80-0572	Det Norske Veritas/Bradford University, 1980
5.	Dexter and Perkins	Component failure rate data with potential applicability to a nuclear fuel reprocessing plant, report DP-1633	El Du Pont de Nemours and Company, USA, 1982
6.	EIREDA	European industry reliability data handbook, Vol. 1, Electrical power plants	EUROSTAT, Paris, 1991
7.	ENI Data Book	ENI reliability data bank—component reliability handbook	Ente Nazionale Indrocarburi (ENI), Milan, 1982
8.	Green and Bourne	Reliability technology	Wiley Interscience, London, 1972
9.	IAEA TECDOC 478	Component reliability data for use in probabilistic safety assessment	International Atomic Energy Agency, Vienna, 1988
10.	IEEE Std 500-1984	IEE guide to the collection and presentation of electrical, electronic sensing component and mechanical equipment reliability data for nuclear power generating stations	Institution of Electrical and Electronic Engineers, New York, 1983
11.	F. P. Lees	Loss prevention in the process industries	Butterworth, London 1980
12.	MIL-HDBK 217E	Military handbook—reliability prediction of electronic equipment, Issue E	US Department of Defense, 1986
13.	NPRD-3	Non-electronic component reliability data handbook—NPRD-3	Reliability Analysis Centre, RADC, New York, 1985
14.	OREDA 84	Offshore reliability data (OREDA) handbook	OREDA, Hovik, Norway, 1984
15.	OREDA 92	Offshow reliability data, 2nd edition	DnV Technica, Norway, 1992
16.	RKS/SKI 85-25	Reliability data book for components in Swedish nuclear power plants	RKS—Nuclear Safety Board of the Swedish Utilities and SKI—Swedish Nuclear Power Inspectorate, 1987
17.	H. A. Rothbart	Mechanical design and systems handbook	McGraw-Hill, 1964
18.	D. J. Smith	Reliability and maintainability in perspective	Macmillan, London, 1985
19.	Smith and Warwick	A survey of defects in pressure vessels in the UK (1962–1978) and its relevance to primary circuits, report SRD R203	AEA Technology—Safety and Reliability Directorate, 1981
20.	WASH 1400	Reactor safety study. An assessment of accident risks in US commercial nuclear power plants, Appendix III, Failure data	US Atomic Energy Commission, 1974

Table 3.5 Component count method

Component	λ	Repair time m	$m \cdot \lambda$
Item 1	λ_1	m_1	$m_1 \cdot \lambda_1$
Item 2	λ_2	m_2	$m_2 \cdot \lambda_2$
etc.			
	$\Sigma \lambda_i$		$\Sigma m_i \cdot \lambda_i$

formal method and the maintainability prediction equation (3.2) is incorporated into the calculation. It is assumed that all components are in series. Any set of elements in parallel should be reduced to an equivalent single element that is in series with the rest of the equipment.

A list of all the components is compiled. For each component, the mean failure rate λ and the repair time m are estimated taking into account environment and stress factors. A set of calculations are then performed as illustrated in Table 3.5.

From the calculations in Table 3.5, the following results are calculated.

(i) The failure rate of the system $= \Sigma \lambda_i$.
(ii) The mean time to failure of the system $= (\Sigma \lambda_i)^{-1}$.
(iii) The repair time $= \dfrac{\Sigma \lambda_i . m_i}{\Sigma \lambda_i}$.

The repair time would usually be the corrective repair time in a design evaluation unless specific allowance has been made for maintenance management considerations.

The use of a table simplifies and organizes the calculations. For large numbers of components, the use of spreadsheet software makes calculation easier.

3.9 Examples of the use of reliability calculations

3.9.1 *Improving reliability by using parallel elements*

There are times when an element in a system has a low reliability compared with others. The element, which is in reality a piece of equipment, may be replaced with another of higher reliability or, as is sometimes the case, an identical element is used in parallel. Suppose we have an element of reliability 0.8 and we wish to improve it by using a number of identical elements in parallel. The reliability of n such elements in parallel is

$$R = 1 - (1 - 0.8)^n$$

For $n = 2$, $R = 0.96$; for $n = 3$, $R = 0.992$ and for $n = 4$, $R = 0.9984$.

There is a significant improvement in reliability for $n = 2$ but after that the improvement reduces. Remembering that the total cost increases by the unit cost of the element (plus installation cost) for each additional item in parallel, then the cost benefit of the high number of units in parallel is not attractive.

If elements in parallel are to be used to improve system reliability, then there is no point putting more elements in parallel than is needed. For example, if the $R = 0.8$ element was in series with four others of $R = 0.9$, $R = 0.94$, $R = 0.95$, and $R = 0.97$, then there would be little justification for using three elements in parallel because the use of two in parallel has already raised the reliability above three of the other elements in the system (unless the $R = 0.8$ element was extremely inexpensive compared with the other elements).

3.9.2 Alternative use of parallel elements

A simple system operates using four elements 1, 2, 3, and 4 which have reliabilities (survival probabilities) of 0.7, 0.75, 0.8 and 0.9, respectively. All elements of the system normally work all the time. The system reliability R_s is

$$R_s = 0.7 \times 0.75 \times 0.8 \times 0.9 = 0.378$$

Assume that the system reliability is considered to be too low and that it has been decided to duplicate the complete system in full active redundancy. There are two ways this can be done. Two series systems can be put in parallel as shown in system 1 in Fig. 3.5 or each element can be duplicated in parallel as shown in system 2 in Fig. 3.5.

For system 1, the system reliability is $R_1 = 1 - (1 - 0.378)^2 = 0.613$.
For system 2, the system reliability is given by

$$R_2 = (1 - (1 - 0.7)^2) \times (1 - (1 - 0.75)^2) \times (1 - (1 - 0.8)^2) \\ \times (1 - (1 - 0.9)^2) = 0.811.$$

Therefore, the more reliable system is obtained by arranging the components as in system 2 without any cost penalty over system 1. (Note that physical constraints may not make this arrangement of components possible in all cases.)

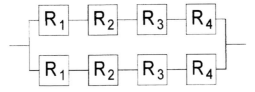

System 1

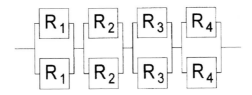

System 2

Fig. 3.5 Alternative system reliability models

3.9.3 Reliability as a function of time

If a component has a mean failure rate of, say, 20×10^{-6} h^{-1} and is assumed constant then its reliability (survival probability) will be given by equation (3.5) as

$$R = e^{-20 \times 10^{-6} \times t}$$

where t is the design life.

For $t = 1$ year (8760 hrs) of operation, $R = 0.839$
  $t = 2$ years of operation, $R = 0.704$
  $t = 5$ years of operation, $R = 0.416$
  $t = 10$ years of operation, $R = 0.17$

The component reliability can be seen to fall exponentially as the design life increases. Therefore, when considering an extended design life, the reliability expectations should be limited accordingly.

3.9.4 A component count example

A machine may be divided into four convenient sub-assemblies A, B, C, and D. The corrective repair times m and nominal mean failure rates λ are given in Table 3.6. The repair times are made on the assumption of good workshop conditions. The machine is to be installed on an exposed site in a chemical plant with reasonable access around the machine. The

Table 3.6 Component count calculations

Sub-assembly	m (h)	λ (h^{-1})	$m\lambda$
A	0.5	5.7×10^{-6}	2.85×10^{-6}
B	1.0	3.5×10^{-6}	3.5×10^{-6}
C	1.2	11.5×10^{-6}	13.8×10^{-6}
D	0.7	10.0×10^{-6}	7.0×10^{-6}
		$\Sigma\lambda = 30.7 \times 10^{-6}$	$\Sigma m\lambda = 27.15 \times 10^{-6}$

machine will be operated at levels up to its nominal rating for short periods but mostly within 80 percent of its capacity. Estimate the mean corrective repair time, the mean failure rate and the MTTF of the machine. The summary of calculations is also given in Table 3.6.

From Table 3.6, the nominal system failure rate $\lambda_0 = 30.7 \times 10^{-6}$ but environmental and stress factors k_1 and k_2 need to be applied. (In this case, the factors can be applied to the overall mean failure rate because the same factors are used for each sub-assembly.) The machine is to be installed on an exposed site in a chemical plant, therefore an environmental factor of 1.5 is applied, see Table 3.2. A stress factor of 0.7 is applied since the machine will be expected to operate usually within 80 percent of its nominal rating with occasional runs up to its normal rating for short periods, this is a subjective interpolation of the factors in Table 3.3.

Using equation (3.15),

$$\lambda = 1.5 \times 0.7 \times 30.7 \times 10^{-6} = 32.2 \times 10^{-6} \text{ h}^{-1}$$

Using equation (3.11)

$$\text{MTTF} = \frac{1}{32.2 \times 10^{-6}} = 31056 \text{ h}$$

Using equation (3.2)

$$M = \frac{\Sigma m\lambda}{\Sigma \lambda} = \frac{27.15}{30.7} = 0.88 \text{ h}.$$

However, some time allowance is required because the machine is located on site where conditions will not be as easy as the workshop. The judgement would have to be made with knowledge of the site but a factor of two might well be applied (or even three or more in winter conditions) giving a mean corrective maintenance time of 1.76 h.

References

(1) Moss, T.R. and **Strutt, J.E.** (1993) Data sources for reliability analysis, *Proc. Instn Mech. Engng, Part E, J. Process Mech. Engng*, **207**, 13–20.
(2) Davidson, J. and **Hunsley, P.** (1994) *The reliability of mechanical systems* (Mechanical Engineering Publications).
(3) Dhillon, B.S. (1983) *Reliability engineering in systems design and operation* (Van Nostrand Reinholt).
(4) Smith, D.J. (1997) *Reliability, maintainability and risk* (Butterworth-Heinemann Limited).

Bibliography

(1) Siddall, J.N. (1972) *Analytical decision making in engineering design* (Prentice-Hall).
(2) Green, A.E. and **Bourne, A.J.** (1972) *Reliability technology* (John Wiley).
(3) Carter, A.D.S. (1986) *Mechanical reliability* (Macmillan, UK).
(4) O'Conner, P.D.T. (1991) *Practical reliability engineering* (John Wiley).
(5) Kelly, A. (1997) *Maintenance strategy* (Butterworth-Heinemann).
(6) Kelly, A. (1997) *Maintenance organization and systems* (Butterworth-Heinemann).

Chapter 4

Design Review

4.1 Introduction

The design review is one of the most important ways of achieving good maintainability and reliability.

The review is much more than the scrutiny of design work in the manner that an examiner checks a student exercise. It should help the designer and enrich design activity. It should be an integral part of design activity and not a 'bolt-on' extra. It should contribute to those problem areas such as maintainability and reliability which may not have been fully taken into account during the search for cost-effective feasible solutions. Evaluation is a normal part of design activity.

The design review may be defined as:

the quantitative and qualitative examination of a proposed design to ensure that it is safe and has optimum performance with respect to maintainability, reliability and those performance variables needed to specify the equipment.

The emphasis on maintainability and reliability does not mean that they should be dealt with in isolation. Wherever possible, all other factors have to be included. Although it is not always possible to do this, nonetheless the principle should be established that maintainability and reliability are an integral part of a comprehensive design review procedure. Availability is not specifically mentioned in the definition. It is a function of mean time to failure and mean repair time. The mean repair time consists of an actual or corrective repair time plus the time to organize manpower, spares provision, etc. The last items are not known at the design stage. Therefore, it is preferable to focus on reliability and maintainability in the review.

In a manufacturing system, variables that specify equipment may refer to production rate, accuracy of assembly, cost, etc. In process plant design, examples of performance include product yield, energy consumption, cost, etc. In equipment or product design, e.g. a car, performance may be the acceleration, top speed, fuel consumption, etc., or for a desktop printer performance may be the quality of the print, the characters printed per second etc.

Safety will have been considered throughout the design process. Mechanisms exist to ensure that certain facilities are as safe as can be expected. Insurance companies have to be satisfied and regulatory authorities examine processes before production can begin. The role of the design review with respect to safety should be to formally record the fact that the appropriate internal and external authorities have been satisfied, that environmental impact factors have been considered and that the necessary safety standards have been adhered to.

4.2 Levels of design

It is useful to undertake a review at four principal levels of design:

(i) design specification review (including market need in product design)
(ii) system review
(iii) equipment (functional unit) evaluation
(iv) component analysis

Nominally, sub-systems would be included in the system level review and sub-assemblies in the equipment review. There is no hard and fast definition of level, equipment in some large systems may be of the scale of a system in another situation, as discussed in Chapter 2. A common sense approach is called for and four main levels of evaluation in a design review will suffice in most cases.

Recognition of level has two uses:

– appropriate review methods can be selected for different tasks, and
– it leads to a systematic approach for an efficient and effective design review.

4.3 A structured design review procedure

A comprehensive design review may therefore be summarized by the following stages.

(i) *Activity*   Review of the design specification (requirements).

   Purpose To ensure that the significance of all the points contained within the design specification is understood.

	Timing	Prior to the commencement of any design activity.
(ii) (a)	*Activity*	Systems level review.
	Purpose	To identify critical areas of the design that may affect plant availability, and to communicate to the detail design teams the necessity to pay particular attention to these areas.
		To comment on the advisability of pursuing projects with a high risk content.
	Timing	Prior to the start of equipment design.
(b)	*Activity*	Systems level review.
	Purpose	To examine equipment groups to maximize uniformity and suitability.
		To maximize the reliability systems formed by manufacturing and process considerations.
	Timing	After the completion of the first equipment designs.
(iii)	*Activity*	Equipment (functional unit) evaluation.
	Purpose	To evaluate quantitatively critical items of equipment. To undertake qualitative reviews of equipment generally.
	Timing	After the completion of the first detail designs.
(iv)	*Activity*	Component analysis.
	Purpose	To check that certain important sets of components will not give rise to maintainability or reliability problems in service.
	Timing	After the completion of the first detail design.

4.4 Design specification

The objective of the design specification review is to ensure that, at the outset, all parts of the specification are understood and that the relative importance of different statements are appreciated. The client and the design team (in-house or contractor) should discuss the salient points in the specification in order to eliminate any misunderstandings. Terms and colloquialisms in common usage in certain industries can sometimes find their way into specifications. Certain terms may well have a particular meaning to the writer of the specification, but the extent of the meaning may be lost to the designer reading the specification. The specification is commonly the reference point in contractual disputes and so it is in the interests of all parties to be clear in the definition of requirements. Contractual matters are discussed in Chapter 8.

Of particular interest in the specification are:

(i) quantified reliability and maintainability objectives
(ii) environmental conditions that may affect maintainability and reliability
(iii) particular maintainability requirements, for example:
  - modular construction (see Chapter 10)
  - workforce maintenance skill level restrictions
  - multi-skill working
(iv) acceptance criteria and demonstration of maintainability and reliability

It is unacceptable simply to refer to maximum reliability and/or maintainability because such statements are meaningless. Maximum maintainability may require that additional components be used, e.g. fasteners to allow access or make certain parts replaceable, which in turn may make the equipment less reliable. Maintainability/reliability trade-offs in equipment design are discussed in Chapter 10. Quantified maintainability and reliability objectives should be written where possible. Design specifications containing maintainability and reliability statements are discussed in Chapter 8.

The specification, which may change during design work, with the client's agreement, is the reference point throughout all design activity and reference to it is the basis of many aspects of the design review. The design team should be encouraged to ask questions at any stage of design if ambiguities or uncertainties arise.

4.5 System review

4.5.1 Prior to detail design

A sensitivity analysis may be conducted concerning overall plant availability. Using flow charts and knowing the nominal production rates of various parts of a plant, buffer capacities, possible operational contingencies etc., it is usually possible to identify critical areas where, if a breakdown occurs, a total plant shutdown may follow after a short time. In some cases, a reliability model of the system may be required to identify critical areas. The objective at this stage is not to undertake a precise quantitative reliability analysis yielding system failure rate predictions because equipment will not yet have been designed/selected.

Systems may have different types of failure, e.g. failure to deliver coolant to a heat exchanger, failure to produce assembled components in an automatic process using adhesives. Fault tree analysis, see Chapter 7,

Section 7.5, determines the relationships between events that give rise to particular types of failure and thus certain critical elements in systems can be identified.

The design review should ensure that the appropriate equipment design teams are made fully aware of the existence of any critical areas that have been so identified. In this way, the design review complements the main design activity by ensuring that extra attention is given to the design of certain equipment.

4.5.2 After equipment has been designed

Design work will be usually carried out based on systems defined by manufacturing or process considerations. There is an opportunity for the design review to complement the initial design by examining equipment groups that 'cut across' conventional system boundaries. For example, a review of the pumps that are to be used in a process plant will reveal if there is an unnecessary diversity of manufacturers that would lead to high spares requirements. A review of 'component transfer equipment' in a manufacturing system will show if there is excessive complexity. For example, in a medium sized assembly line it was found that a number of pneumatic component transfer mechanisms had been designed on different principles requiring a wide variety of components and different set-up and adjustment techniques. The resulting complexity would have been expensive to maintain and prove unreliable in service. Therefore, using experience, equipment groups should be defined and analysed to maximize uniformity to reduce spares. If diverse products are avoided, maintenance teams can more readily build up knowledge and competence in maintenance practice.

After the detail design stage, it is possible to carry out reliability analyses of systems defined on process or manufacturing bases since the details of the equipment in each system will be known. Reliability models can be used at this stage to determine if reliability objectives have been achieved. There is an iterative interaction between the systems and equipment levels of the design review procedure as shown in Fig. 4.1.

Systems evaluations of a more general kind than reliability analysis can be performed:

(i) The parameter profile analysis method (see Chapter 6) may be used to review systems to determine their character and, importantly, highlight weak areas with respect to system performance including maintainability and reliability.

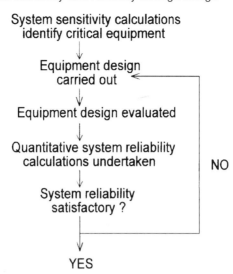

Fig. 4.1 Interaction between system and equipment levels in a design review

(ii) Failure mode analysis (see Chapter 7) can be used to find which parts of the system are critical from the combined perspectives of the probability of failure and the consequence of failure.

4.6 Equipment evaluation

Equipment evaluation is an important part of a design review. Different items of equipment require different evaluation techniques. It is at the equipment level that the engineer has the opportunity to embark upon a quantitative appraisal since firm proposals have been set down and specific hardware proposed.

The system reviews may have identified certain critical items of equipment or weak spots that require in depth evaluation. Different evaluation methods may be employed:

(i) A method that is particularly suitable for concept evaluation is given in Chapter 5.
(ii) The device performance index (Chapter 5, Section 5.6.5) is a method that evaluates equipment that has been designed in detail or compares alternative proposals in a selection/procurement exercise. The method combines quantitative assessments with respect to different performance parameters including maintainability and reliability and is able to incorporate subjective value judgements.

(iii) The parameter profile analysis (Chapter 6) method evaluates equipment performance as well as system characteristics.
(iv) Failure mode and maintainability analysis (Chapter 7, Section 7.2) may be used to evaluate repair actions and to specify appropriate condition monitoring methods.

Chapter 5 deals with equipment evaluation methods in depth including the use of check lists.

The design review should assess the risks associated with novel designs and new technology. Reliance on unproven technology in a critical area may be unwise. Evaluation may take the form of an appraisal by a research and development department.

Many items of equipment, including common group sets of equipment, do not require in-depth evaluation. It is impractical to carry out such an undertaking on a large system in any case. Check lists provide a powerful and efficient way of ensuring that equipment is fit for purpose, see Chapter 5.

Maintenance methods such as condition monitoring should be considered in equipment design. Provision for condition monitoring is often made easier at the design stage rather than by retro-fitting.

The choice of evaluation method is made according to the case under consideration. It is better to integrate the evaluation with the design activity rather than to undertake a series of evaluations by 'outsiders' after design work has been completed. The latter approach gives the feeling of examination of design ability and is not constructive.

4.7 Component analysis

The demarcation between equipment evaluation and component analysis is not precise; the question is one of scale. In the following discussion, a component will be an item that is a replaceable part. It may be: a bearing in a special machine; a gasket; an electric motor; process valve etc.

It is clearly impractical to consider a general survey of components in a manufacturing plant. Some cases for detailed scrutiny will naturally follow from the evaluation at equipment level. Certain component classes may be identified for detailed analysis where there are particularly demanding duties, for example, the seals in a complete process line carrying a corrosive fluid. Experience is important to identify component sets for analysis.

Chapter 5 discusses component analysis in detail and includes:

- check lists
- methods for components handling fluids
- comparative reliability analysis of components, and
- analytical approaches

4.8 The design review team

4.8.1 Role of the review team

The design review activities should be integrated as far as possible with normal design activities. This is an important principle. Therefore, specific evaluation exercises are best performed by the designers who have carried out the original design work. Evaluation should be a normal part of design activity. The concept of a design review team that carries out (or causes to be carried out) evaluations that check up on the work of designers is not constructive.

That said, there is a role for a design review team:

(i) to ensure that the initial design specifications are fully understood
(ii) to comment on and to communicate the results of the system review to equipment design teams regarding critical areas
(iii) to review the results of the in-depth equipment evaluations
(iv) to define common equipment groups for review
(v) to identify component sets for evaluation based on experience
(vi) to provide expert opinion on maintainability and reliability to aid design teams

4.8.2 Composition of the review team

In order to carry out a design review effectively the team must be multi-disciplinary in character. Also, specialist expertise will be needed to undertake detailed studies. Taking into account all the above discussions, the following team composition is suggested.

(i) *Design (A)* There must be a direct link with the original design team in order that comments may be made on the implications and feasibility of suggested modifications.
(ii) *Design (B)* A designer with relevant experience, but not involved in the design under review, should be available to provide a fresh perspective.
(iii) *Research and development* Specialist expertise will be necessary to give advice in certain areas (e.g., corrosion science, vibration analysis). A person with broad experience is required who can

recognize the need to co-opt a specialist or to commission detailed studies as necessary.

(iv) *Maintenance* The design team should have had feedback of maintenance and reliability field data if they are available. However, maintenance experience on the review team will provide firsthand knowledge and should be capable of identifying problems associated with feasibility of operation, staffing levels, demarcation, etc.

(v) *Production management* Experience of similar plant is always useful. Staffing levels, industrial relations, and management efficiency in the new plant are relevant.

(vi) *Safety* Problems associated with personnel and environmental safety are always important. A safety specialist may not always be required, but should see the proposals at every stage.

(vii) *Project engineer* The project engineer is the link between design, commissioning, and production. The project engineer also will be familiar with most aspects of the design since she/he is responsible for the bringing the project to fruition. The project engineer, therefore, is in the best position to lead the design review team.

The number on the review team is therefore between six and nine depending upon the topic under discussion. This is a manageable group. The above list of personnel is not intended to be definitive but rather is given as an indication of size and expertise.

The question of whether a design review should be undertaken by personnel outside the client company must be considered. No doubt there is an argument that 'fresh minds' should bear on the subject and, therefore, an 'outside team' should be employed. However, the strongest argument surely is that the personnel who will be ultimately responsible for operating the plant should be heavily involved in the reviewing procedure. After all, their livelihoods will depend upon the plant long after the review team has been disbanded. Also, the presence of the extra designer not connected with the design under examination provides a fresh mind. These arguments weigh in favour of an 'in-house' review team, but a compromise situation is possible whereby the extra designer and the research and development expertise is provided by a specialist organization. The latter contribution could only be justified if the operating company did not already support a research and development department of its own.

4.8.3 *Management and control*

The review team effectively manages the evaluation of the design from the review of the design specification through to detailed component checks. It is the responsibility of the review team to ensure that, within reason, potentially serious operating difficulties with the plant are avoided. The responsibility for decisions concerning the acceptance of design proposals which have been subjected to particular examination by the design review team rests with the team.

Specific analysis techniques which are applied to systems, equipment, and components form integral parts of the review procedure. These analyses are best carried out by design team as part of normal design activity. However, the review team retains its overall control of the design evaluation without becoming submerged in detailed calculations. The review team must, however, understand the principles of an analysis methods in order to properly interpret the findings.

Experience within the review team is important. For the case of the system review prior to detail design, the team has to comment on the relative risks associated with any novel designs. Attitude to risk is dependent upon the operating environment of the company, both in the workplace and the business climate and, therefore, experience within the company is required.

4.9 General discussion

It is important that sufficient time is allocated for a design review when planning a project programme. Sufficient emphasis should be placed on the review in the early stages of design, for it is wasteful to spend large amounts of time and money correcting mistakes which arise because of poor basic design principles or lack of appreciation of the business need. Good foresight and a sound design foundation will undoubtedly save money in the long term. There is no reason to presume that the use of a systematic design review will lengthen the total time to achieve a working system or product. The longer design time should be (possibly more than) compensated for by a reduction in product development or plant commissioning time. A further pay-off will also be in the form of good maintainability, reliability and performance.

Sometimes there will be pressures to short cut design review procedures. If there have been delays, say in some design work, there may well be moves to cut out parts of a design review to same time. Such approaches are short sighted and, if taken, may well lead to pre-production start up difficulties or poor initial product performance. One of the advantages of integrating design evaluations with original design activities is that they are not seen as an 'add-on extra' that can be cut for short term gain.

It is possible to extend the design review procedure to post manufacture and use by defining two further stages:

(i) performance comparison with design
(ii) decommissioning inspection compared with design

The performance comparison with design and decommissioning inspection are sources of design information. Feedback to design is dealt with in Chapter 13. Inspection on decommissioning is also a source of design information and applies to particular industries. In addition to information gathering, the review should ask higher level questions concerning how well the design objectives have been achieved and what lessons can be learned about how to approach the design of such systems in future, e.g. how a large project was sub-divided, maintenance and reliability knowledge of sub-contractors, problems at interfaces between systems.

Finally, it is interesting to note the trends in design review activities. Engineers have always checked design work therefore the idea of a 'design review', sometimes known as a 'design audit', is not new. In the 1970s and 1980s, there was much interest in systematic approaches to design reviews in order to achieve good quality design work (references (**1**)–(**8**)). Such work helped to raise the awareness of the importance of design reviews and many companies use design review procedures. Today, design reviews are used to varying degrees and some candid personal comments from professional engineers on their experiences reveal interesting trends.

In a number of discussions with engineers, it is apparent that the design review meetings have become progress meetings in many cases. Little design evaluation is done. Most discussion concerns the completion of the design work on time. Another comment frequently made, especially by maintenance and production engineering managers, is that their companies have design review procedures that involve them but that their role is ineffectual. Although they are given an opportunity to comment on completed design work, especially design work carried out under contract, there is insufficient time to review the work effectively. They comment that modification to completed design work is an unrealistic option and that the project cannot usually be held up for a sufficient length of time to allow them to make a significant contribution at such a late stage. Therefore, for a number of reasons, the existence of design reviews may satisfy an audit of company procedures and quality systems but may not be doing much good.

References

(1) Jefferies, B., Sadler, J. and **Williams, P.R.** (1980) The design audit, *Terotechnica* **1**, 237–241.

(2) Jefferies, B. and **Westgarth, D.** (1981) The design audit, *CME*, **28**, 24–26.

(3) Baker, G.J. (1981) Quality assurance in engineering design, *Engineering Designer*, **3**, 16–18.

(4) Napier, M.A. (1979) Design review as a means of achieving quality in engineering design, *Engineering*, **219**, 87–91.

(5) Engineering Design Handbook, US Army Material Command, AD823539.

(6) Levesque, C.R. (1978) Design audit concept in new product development, ASME paper No 78-DE-W-2, pp. 1–4.

(7) Jacobs, R.M. and **Mihalaski, J.** (1973) Minimizing hazards in new product development (ASME) pp. 45–55.

(8) Thompson, G. (1985) *Design review* (Mechanical Engineering Publications).

Chapter 5

Equipment Evaluation

5.1 Introduction

Evaluation is a major part of design activity. It is a decision making activity. Original design work is evaluated at the concept stage and when the details of a machine have been worked out. Different evaluation methods are required for different activities. Methods used for assessing detail design are generally inappropriate for the evaluation of design concepts because the details of the equipment are not yet available at the concept stage. In many cases, proprietary equipment will be selected and incorporated into larger design schemes therefore methods for comparing equipment are needed. Small items of proprietary equipment can be evaluated in great detail, but to carry out very detailed assessments of large machines would be extremely time consuming and in most cases inappropriate. Also, there will be times when only a limited amount of detailed information is available about some proprietary equipment. Therefore, the evaluation of equipment needs to be considered with respect to the particular case under consideration and appropriate methods adopted.

Maintainability and reliability considerations can be included very effectively into equipment evaluation. They can be integrated with other performance parameters such as cost, weight, and production output and quality. Alternatively, maintainability and reliability can be considered individually if required. Qualitative and quantitative methods may be used according to the requirements of individual cases.

Many engineers do not rely on formal design evaluation methods but tend to use their own judgement. A 'black box' approach like this may suit a minority but many poor decisions are made this way. Interestingly, discussions with the few who make good 'black box' decisions reveal in them a clarity of thought that embodies the principles of evaluation and decision making that underpin formal methods.

In this chapter, a range of evaluation methods and principles are given. They include methods suitable for concept and for detail design evaluation, for the selection of proprietary equipment, and for use in qualitative and quantitative assessments. An indication is given of their most suitable application but an open mind is needed in design evaluation. It may be that particular circumstances pertain that could well suit the use of a method in a case for which is is not normally used. Importantly, knowledge of the principles of the evaluation methods described will enrich any future 'black box' design evaluation and decision making activities.

5.2 Check list

5.2.1 A simple check list

Check lists are a simple but effective way of equipment evaluation. They are usually made up of a series of short questions that direct the attention of the analyst to some particular aspect of the design. The response to each question is usually YES/NO and on completion of the review the equipment is deemed either acceptable or not. Of course, further enquiries concerning some of the adverse responses may be required before accepting equipment.

A check list evaluation may be made using assembly drawings or on manufacturers' equipment. A detailed check list is not suitable for concept design evaluation.

The following list contains questions that are particularly relevant to maintainability and reliability. It is not an exhaustive set of questions and additional questions should be added to suit individual circumstances. The questions have been phrased so that an answer YES is a favourable response. This makes a review of the check list easier as one looks for NOs.

Accessibility:	Is there reasonable access to components for fault finding and parts replacement?
Fasteners:	Can maintenance tools be located on fasteners? Is there a minimum diversity of fastener types used?
Adjustments:	Are there satisfactory mechanisms/devices provided for any fine adjustments? Is the degree of adjustment measurable? Can adjustments be made before production starts?

Construction:	Is the equipment made of sets of parts in modules to facilitate replacement?
	Are conventional tools adequate to perform maintenance tasks?
Simplicity:	Does the equipment give the impression of being simple and not over-complicated?
	Does the equipment look robust?
Spares:	Is the variety of spares required reasonable and not excessive?
	Is the future availability of spares assured?
Personnel:	Are normal skill levels required for maintenance procedures?
	Are the numbers of trades required for maintenance a minimum?
Ergonomics:	Can the forces/torques required for maintenance be provided by persons of average physique?
Faults:	Are people safe in the event of maloperation?
	Is other equipment protected in the event of maloperation?
	Can the operator readily detect if the machine operates out of specification?
Condition monitoring:	Is provision made for hand-held condition monitoring devices to be used?
	Is provision made for installed condition monitoring instrumentation if required?
Corrosion:	Are the components, and especially fasteners, resistant to external corrosion?
	Are the materials selected to resist the internal corrosion of any parts handling fluids?
Technology:	Is the design of the equipment based on known technology?
	Is the confidence level in any novel technology adequate?
Manufacturer:	Has the manufacturer a good reputation?
	Is a quality assurance system used?
Cost:	Is the cost reasonable? (excessively cheap equipment may be unreliable)

5.2.2 *The use of check list to compare equipment*

Check lists may be used to compare equipment by different manufacturers. Instead of a simple YES/NO response, a score can be given on a simple 1, 2, 3, 4 scale for each question. The overall score indicates the more attractive equipment. (Addition may not be the best way to combine scores; methods to derive an overall score are discussed in Section 5.6.5 below.)

Check lists may be developed in more depth using computer software. A 'pull-down' help menu can provide supporting information if a question needs elaboration. The use of well designed software also helps present lots of information clearly and simply when required without overfacing the user. Also, any quantitative scoring of responses is done immediately and automatically and weak areas can be visually highlighted in the software.

5.2.3 Advantages of a check list
The advantage of a check list is that it focuses attention on salient aspects of equipment design and a more thorough and meaningful evaluation is obtained than if one peruses a design in a casual manner.

The use of a check list has an important spin-off in areas outside the scope of the questions on the list. During the close scrutiny of the design, weak aspects of the design very often come to light even though they have not been covered by a question. This is because the design is being reviewed systematically and the features and principles of the design need to be understood in order to answer questions.

5.3 General discussion about equipment selection

5.3.1 Costs
Most equipment selection is made on purchase price. Even though the life cycle costs of equipment may be recognized as being more important than initial cost by many engineers, they are often constrained by a requirement to reduce initial costs. A thorough selection process can provide useful arguments to choose the most appropriate long term cost-effective solution.

There are many factors that affect the choice of equipment. In this discussion, factors that affect maintainability and reliability will be considered. Other performance parameters will vary according to each case. From the perspective of the equipment supplier, maintainability and reliability can be difficult topics. For example, high reliability is a useful selling point, but there is a limit to how much costs can be increased to achieve high reliability before sales will suffer.

5.3.2 Component quality

The reliability of equipment is highly dependent on the quality of the components used. If good quality components have been used then this will be reflected in the selling price. Products that are offered at a cheap rate should be treated warily. However, cheaper components could be used and the selling price held high to maximize profit so, in general, enquiries about components should be made for critical equipment. There are certain components that can be used as reliability indicators, for example:

- rolling element bearings
- hydraulic components
- pipe couplings
- pneumatic equipment
- gaskets

There are many such components that are bought in items. Equipment manufacturers should be able to confirm that components from reputable component manufacturers have been used. Unbranded products can be used to cut cost but there may be a reliability penalty.

Quality control in the design and manufacture of equipment is a positive contribution to the production of reliable equipment. Adherence to a quality standard does not necessarily mean that equipment is reliable, but not having a quality control of materials and manufacturing process will almost certainly mean unreliable equipment.

5.3.3 Spares, servicing and technology

Medium/long term provision for spares should be assured. If many non-standard components are used then spares may be expensive. Of course, standard spares may be obtained from a variety of sources, therefore reliance need not necessarily be placed on the original equipment supplier. Service contracts may be entered into from the outset or service may be sought as and when required. Whatever arrangement is preferred, the provision of technical support in the future has to be considered. This is especially important if novel technology is involved. Whilst the benefits of novel technology may appear attractive, the prospects of dealing with problems during production may be daunting. The maintenance workforce may not be familiar with equipment and external support may be required frequently and/or urgently at a critical time if equipment proves troublesome. A service contract for equipment based on novel technology could well be advantageous.

5.3.4 *Operating environment*

Enquiries should be made about the use of equipment in similar operating environments. The reliability of equipment is highly dependent on the environment therefore reliability claims should be treated with caution until supporting applications are provided or calculations carried out. The response of the manufacturer to reliability and maintainability enquiries is a useful indicator. It is reassuring if the manufacturer has a general awareness of maintainability and reliability matters. Exaggerated claims should raise alarm. If it is evident that the manufacturer does not know much about reliability then there is need for caution. For example, an enquiry about reliability to a manufacturer should probably be met with questions about the intended use of the equipment, operating environment and loading before reliability statements are made.

5.3.5 *Condition monitoring*

Condition monitoring may be an integrated part of equipment, in which case the design should incorporate the required instrumentation. In other cases, careful consideration should be given to the subject. The principal failure modes should be identified and a check made to determine how these may be detected. The objective is to look ahead and try to anticipate problems. If a significant failure mode that could cause serious production problems cannot be identified then modifications should be insisted upon before accepting the equipment.

Many operating companies use routine condition equipment such as hand-held bearing vibration detectors, temperature monitors, thermal imaging etc. (For a review of condition monitoring techniques see Appendix 1.) Equipment should be reviewed to determine if such instrumentation can be placed in appropriate places. It is frustrating and costly if equipment has to be modified at a later stage to allow condition monitoring devices to be used.

5.3.6 *Equipment design features*

Modular construction greatly eases maintenance. If components are designed in sets for ease of replacement then fast maintenance times are possible. General good accessibility, the use of sensible fasteners etc. are all indicators of an awareness of maintainability by the original equipment designer. The check list in Section 5.2.1 will identify problems in detail design. Chapter 10 gives examples of detail design to achieve good maintainability. With these examples in mind, the review of equipment design will reveal whether or not maintainability has featured in the considerations of the designer.

Equipment Evaluation 57

5.3.7 *A qualitative systematic procedure*
A systematic qualitative evaluation of equipment can therefore be undertaken with respect to maintainability with reference to:

(i) component quality and cost
(ii) spares, servicing and technology
(iii) operating environment and reliability
(iv) condition monitoring
(v) equipment design features using the detailed check list review of the engineering features of the equipment

5.4 Comparative reliability analysis

5.4.1 *Method*
Reliability modelling may be used to compare two or more items of equipment. The method of analysis is as follows.

(i) For a defined mode of failure, examine the design and list all the events that may lead to failure. For example, in a pump, one failure mode is 'failure to contain fluid' and one of the failure events that would lead to that failure mode is leakage through the shaft seal.
(ii) Compose an event structure based on the failure events using series and parallel elements as required to model the reliability.
(iii) Analyse the event structure using conventional statistical methods letting each event have a reliability R_i which is unquantified.
(iv) Repeat this process for alternative designs and compare the results.

There is no requirement for statistical data. Instead, a comparison between different designs is made on a relative basis. When a different item in a component can be specified to perform the same function, e.g. bellows sealing rather than stem packing in a valve, then personal judgement is used to compare the relative merits of each method. The reliability of one component is adjusted to a fraction of its counterpart in the competing design to allow the comparison of the designs to proceed.

5.4.2 *Example: Comparative reliability analysis of two process valves*
The objective of this analysis is to compare the reliability of two valves with respect to leakage. They are both globe valves but they have different internal features concerning their sealing arrangements:

Valve A, see Fig. 5.1(a), has a replaceable seat and the stem is sealed by a metal bellows plus stem packing as a safeguard should the bellows fail.

Valve B, see Fig. 5.1(b), has an integral seat and the stem is sealed by two sets of stem packing.

The failure mode considered is 'Either leakage of fluid through the valve or out of the valve'.

Valve A

The leak paths for valve A are:

(i) bonnet/body gasket failure
(ii) valve seat/body gasket failure
(iii) disc/seat leakage
(iv) bellows failure, and
(v) stem packing failure

Let the reliabilities of the elements (i) to (v) be denoted by R_1 to R_5, respectively. Elements (iv) and (v) form a parallel pair since both must fail for the system to fail. Full active redundancy is assumed since the packing seal is always being worn by the motion of the stem even though it is not under pressure. Elements (i), (ii), (iii) must all survive for the system to survive, therefore the reliability model of the valve is as shown in Fig. 5.2.

The reliability of valve A is then:

$$R_A = R_1 \times R_2 \times R_3 \times (R_5 + R_4 - R_4 \times R_5) \tag{5.1}$$

Valve B

Valve B has an integral seat, therefore there is no valve seat/body gasket to fail ($R_2 = 1$). The stem seal is by two sets of stem packing therefore there are two elements of R_5 in parallel. The reliability of valve B is therefore:

$$R_B = R_1 \times R_3 \times (2 \times R_5 - R_5^2) \tag{5.2}$$

In order to compare the two designs a judgement concerning the relative merits of bellows sealing and stem packing must be made. A realistic assessment would suggest that the bellows is much less likely to leak, say $R_4 = 2 \times R_5$.

Therefore equation (5.1) becomes

$$R_A = R_1 \times R_2 \times R_3 \times (3 \times R_5 - 2 \times R_5^2) \tag{5.3}$$

Dividing equation (5.3) by equation (5.2) gives

$$\frac{R_A}{R_B} = \frac{R_2 \times (3 - 2 \times R_5)}{(2 - R_5)}$$

Equipment Evaluation 59

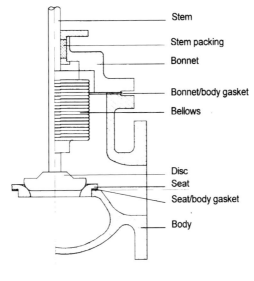

(a)

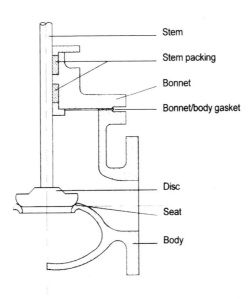

(b)

Fig. 5.1 Alternative designs of valve

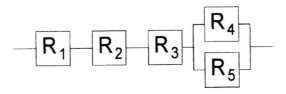

Fig. 5.2 Valve reliability model

Now, the maximum possible value for R_5 is 0.5, since it has been decided that $R_4 = 2 \times R_5$ and R_4 cannot exceed 1. The integrity of a bellows seal is excellent, so that whilst R_4 cannot strictly be unity, a reasonable estimate for R_5 is 0.5.

Substituting $R_5 = 0.5$ gives

$$\frac{R_A}{R_B} = \frac{4}{3} \times R_2$$

For valve A to be more realistic than valve B, the reliability of the seat gasket (R_2) must be greater than 0.75, therefore the conclusion focuses on the reliability of the valve seat/body gasket.

Let the mean failure rate of the gasket be λ and the valve design life be t. Assuming constant failure rate and using equation (3.9), we require

$$0.75 < e^{-\lambda \times t}$$

giving

$$\lambda > -\frac{\ln(0.75)}{t}$$

Therefore, for the required design life t the minimum mean failure rate can be found. For $t = 131400$ h (15 years), $\lambda_{min} = 2.2 \times 10^{-6}$ h^{-1}.

Therefore, in design A, if a gasket or other seal is used for the seat/body seal that is expected to have a mean failure rate greater than 2.2×10^{-6} h^{-1} then valve A will be more reliable. The decision on λ_{min} has to be made with respect to the fluid and the operating conditions.

5.5 Evaluation of design concepts

A very useful method of evaluation has been proposed by Pugh, reference (**1**), which is useful for concept evaluation. It is similar to the 'Paired Comparison' method of evaluation used in Creative Problem Solving, see reference (**2**). The method is a qualitative evaluation in which design concepts are compared to a reference design concept. The reference concept may be a 'standard' design, some equipment that is presently used, a design that is considered just acceptable or one of the proposed concepts that appears to be the favourite on first inspection.

An evaluation matrix is constructed as shown in Fig. 5.3(a), comprising the concepts 1 to m which are arranged against the assessment criteria 1 to n. A brief sketch of the concepts should be included on the matrix to illustrate 1 to m if possible. The reference concept is chosen as the datum. Each concept is then compared with the datum with respect to each assessment criterion independently. If a concept is better than the datum

Concepts vs Criteria Matrices

(a) Concept evaluation matrix (empty): Criteria 1..n vs Concepts 1 2 3 4 ... m, with "DATUM" written across the body.

(b) Completed concept evaluation matrix:

Criteria	1	2	3	4	... m
1	−	+	D	+	S
2	+	+	A	−	−
3	+	S	T	−	S
4	S	−	U	−	+
n	−	S	M	S	S
Total +	2	2		1	1
Total S	1	2		1	3
Total −	2	1		3	1

Fig. 5.3 (a) Concept evaluation matrix. (b) Completed concept evaluation matrix

with respect to a certain criterion then a + is inserted into the matrix for that concept against that criterion. If the concept is worse than the datum then a − is inserted and if it is the same as the datum or if no judgement can be made then an S is inserted. Thus the matrix is completed with +, − and S points and the +, −, and S ratings are totalled for each concept. Figure 5.3(b) shows the appearance of a completed matrix.

The matrix highlights the strengths and weaknesses of concepts. The objective of the evaluation process is to eliminate weak concepts and to identify those strong concepts that are suitable for further design work. The − and S points of the concepts are reviewed to see if significant improvements can be made. In Fig. 5.3(b) concepts 1 and 2 appear strong and concept 4 is weak. Concepts 1 and 2 would be reviewed with respect to their weak criteria to see if their rating could be improved to + before the final choice between these two concepts is made. In this way concepts can be systematically reviewed to make them robust and suitable for further design work.

The method is suited to evaluations with respect to maintainability and reliability. Assessment criteria that pertain specifically to maintainability and reliability are defined in terms that are suitable for concept evaluation. For example, in the case of maintainability a repair time would be inappropriate because the repair time could not be calculated from the information available at the concept stage. Similarly, mean time to failure or mean failure rate calculations cannot be made. Instead, criteria should be used that refer to such factors as:

- simplicity and elegance of the design
- minimum number of parts
- suitability for modular construction
- technological uncertainty and the use of new technology, and
- unavoidable requirement for highly stressed parts

Also there are specific operating and environmental criteria that pertain to reliability which can be included. For example, in the case of mobile equipment, the requirement for a minimum number of trailing cables would lead to high reliability. Operating and environmental criteria are case-specific and they should be added to the general maintainability and reliability criteria given immediately above.

Chapter 9 shows how this evaluation method can be integrated into concept design and an example is given of the reliability factors that were used in a concept design study.

5.6 Systematic quantitative equipment evaluation

5.6.1 Introduction

Systematic quantitative evaluation of equipment has several uses:

(i) For the purchase of large equipment, to focus attention on the salient features of the equipment and ensure fitness for purpose.
(ii) When alternative proprietary products are being considered, to provide a clear comparison to enable a choice to be made.
(iii) In detail design, to provide a thorough analysis of equipment that system studies have identified as critical with respect to maintainability, reliability and other performance.
(iv) The use of an open, transparent method of analysis may be used to justify not choosing the cheapest option. There are many cases when it is understood by engineering staff that the choice of the equipment based on minimum purchase price will probably lead to reliability and/or maintainability problems.

The method is not generally suitable for concept design evaluation. This is because the detailed information is not usually present at the concept stage to enable calculations to be performed in order to undertake a systematic quantitative evaluation of equipment.

Maintainability, reliability and those performance criteria that are needed to specify the equipment are all included in the evaluation. The performances of the equipment with respect to all the criteria are combined to give an overall performance. In this way, the usefulness of an item of equipment to the system of which it is a part is found.

5.6.2 Basic principles
A quantitative evaluation has three components:

(i) assessment criteria
(ii) measurement
(iii) value

The criteria against which any design proposal is to be assessed should be clearly specified. The assessment criteria should be defined so that the performance of a design can be assessed against each criterion by calculation. Vague or ambiguous criteria are of little use in quantitative design evaluations. In the case of maintainability, mean corrective repair time would be an appropriate criterion. For reliability, the mean time to failure or mean failure rate would be appropriate.

Calculations are then performed using the available design information to predict the performance of the design with respect to each criterion. Precise calculations should be carried out with as great an accuracy as possible.

Having defined assessment criteria and performed calculations with respect to the criteria, the value of the calculated results has to be assessed. There is a subjective element to design evaluation exercises. For example, a mean corrective repair time of 0.75 h may be considered good by some but mediocre by others, even in the same organization. A mean failure rate of, say, 450×10^{-6} h^{-1} is meaningless unless referred to a particular application. Also, each assessment criterion may not have the same importance; some may consider reliability more important than maintainability or output rate may be less significant than the quality of the output. Therefore, the calculated results have to be evaluated against a specified set of value judgements.

5.6.3 A systematic quantitative method
The basic assessment method is:

(i) *Define the assessment criteria*
Use the minimum number of criteria that will suffice. Criteria should be independent.
(ii) *Set the value judgements*
For each criterion, define a range of performance from an upper value of perfectly acceptable performance to a lower limit that

defines the threshold of complete unacceptability. Most engineers have an expectation of equipment performance. All that is required here is to set the bounds of expectation in terms of an excellent performance down to an unacceptable performance. A value score of 10 is given to the perfectly acceptable performance (and all performances better than this). A value score of 0 is given to the lower limit that defines unacceptable performance. Value scores of performances between the two extremes are determined by interpolation using a linear utility function. Figure 5.4 shows utility functions for the mean time to failure and mean time to repair for a particular application. Note that the value judgements are made for the application.

(iii) *Determine the relative importance of criteria*

Assign a scale factor k_i for each criterion to describe its importance. Use 1 as the minimum and increase to represent increased importance, 2 = twice as important. Only use scale factors if there is a very good reason to differentiate between criteria, otherwise equally weight all criteria (all $k_i = 1$).

(iv) *Performance prediction*

Consider the proposed design and, using calculations, predict its performance with respect to the assessment criteria. In the case of proprietary equipment selection, the manufacturers would be asked to supply the required information.

(v) *Convert the performances to value scores*

Convert the performance calculations to value scores u_i using the utility functions, see (ii) above.

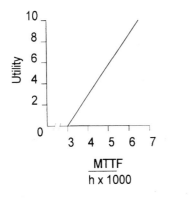
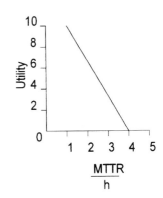

Fig. 5.4 Examples of utility functions

(vi) *Overall value*

The overall performance value P of each proposal is then calculated as

$$P = u_1 \times k_1 + u_2 \times k_2 + u_3 \times k_3 + \cdots u_n \times k_n \tag{5.4}$$

where k_i = scale factors for each criterion
u_i = performance score for each criterion from the utility function
n = number of criteria.

The design proposal with P = maximum is the preferred choice.

5.6.4 Comments on the method

The use of addition to obtain an overall score is favoured by some engineers for its simplicity. However, if many criteria are used then there is a strong possibility that a poor score with respect to one variable may be compensated for by a set of good scores with respect to other variables. Thus, a proposed design could be rated highly even though it is almost completely unreliable. This drawback may be overcome by insisting that each criterion should have a score greater than zero, i.e. each criterion should have a score in the acceptable range. If any proposal has a projected performance below zero for any criterion then that proposal is rejected.

The number of criteria used should be limited. In most engineering cases, there are a relatively small number of factors that effectively govern the behaviour of equipment. Usually it is less than six. An important aspect of the analysis method is the careful selection of criteria. The use of many criteria means that the analysis process becomes burdensome and attention is distracted from the salient features of the design. In any case, if many criteria are included in the analysis, there is a tendency to produce overall performances of a similar level for all designs because the good and bad aspects of each design tend to average out to a common level by the addition process. Therefore, it is often difficult to choose between proposals using this method if six or more criteria are used.

A linear utility function is used to convert performances into value scores. This process is based on sound engineering judgement concerning the limits of performance from excellent to unacceptable. *A value score should never be given to a performance calculation without such a clear justification and under no circumstances should a score be simply 'plucked from the air'.*

Scale factors can significantly change the outcome of this assessment method. They should only be used if there is a strong reason. In general it is better to start from the basis of not using scale factors at all, i.e. all

$k_i = 1$, and then to consider their use only if justification becomes apparent during the analysis.

5.6.5 Device performance index

Device performance index (*DPI*) is an improvement over the addition method of obtaining an overall score. It is based on an inverse method of combining individual value scores. Otherwise the analysis method is identical to the method described in Section 5.6.3 concerning the definition of criteria, setting value judgements, performing calculations and obtaining value scores.

The overall value is found by calculating the *DPI* as follows.

$$DPI = n \times \left(\frac{1}{u_1} + \frac{1}{u_2} + \frac{1}{u_3} + \cdots \frac{1}{u_n} \right)^{-1} \tag{5.5}$$

where u_i = the value scores for each criterion.
n = the number of criteria.

The use of n in equation (5.5) brings the *DPI* calculation conveniently into the range 0–10, which is the range of the value scores, no matter how many criteria are used which makes evaluation easier.

If scale factors are required, then

$$DPI_k = U \times K \tag{5.6}$$

where

$$\frac{1}{U} = \frac{k_1}{u_1} + \frac{k_2}{u_2} + \frac{k_3}{u_3} + \cdots \frac{k_n}{u_n}$$

and

$$K = k_1 + k_2 + k_3 + \cdots k_n$$

The device performance index method has a principal advantage over simple addition because low performances with respect to one criterion cannot be compensated for by other high scores. For example, for three criteria, the *DPI* is

$$DPI = \frac{u_1 \times u_2 \times u_3}{u_1 \times u_2 + u_2 \times u_3 + u_1 \times u_3} \times n$$

If there is a low score with respect to one criterion, then the numerator will also be small because it is the multiple of all the individual value scores. Therefore, if a design scores very low with respect to maintainability then the overall *DPI* will be small. (Note that the denominator in the *DPI* calculation makes the calculation less sensitive to errors in u_i than simple multiplication, which is difficult to use with scale factors in any case.)

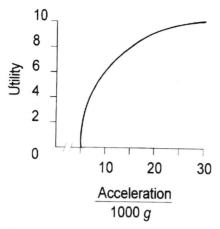

Fig. 5.5 Utility function for acceleration

5.6.6 *Non-linear utility functions*
A non-linear utility function may be used if appropriate. Figure 5.5 shows a utility function for the centripetal acceleration of a centrifuge. The centrifuge is considered perfectly satisfactory if it develops 3×10^4 g. If the acceleration is only 1×10^4 g then the centrifuge is still considered reasonable with a score of 6, but below 1×10^4 g the performance is considered to be poor. Such a utility function would describe a centrifuge application where an acceleration of the order of 1.5×10^4 g is needed to separate the contamination to a reasonable level of purity; above this more acceleration is of limited benefit; below this value the product may not be cleaned well.

Utility functions may be derived by careful consideration and graph sketching as the analyst peruses the problem. Alternatively, the analyst may conduct interviews with suitably experienced engineers. Typically, the interviewee is asked to score, out of ten, a range of performances that are expected of a machine in a defined working environment. (Note that it is the performance which is scored, not any particular machine.) By considering a range of performances, points on the utility function are obtained. During interviews, it is useful to work in from the extreme values, repeating questions to verify points. The utility function is the sketched to best fit the points recorded. Figures 5.6 and 5.7 show examples of utility functions that have been obtained by the interview process.

5.6.7 *Illustrative example of a device performance index calculation*
Assume that three manufacturers have been approached concerning the supply of a machine for a production process. It has been decided that the machines should be judged on four criteria:

(i) maintainability (mean corrective maintenance time)

68 Improving maintainability and reliability through design

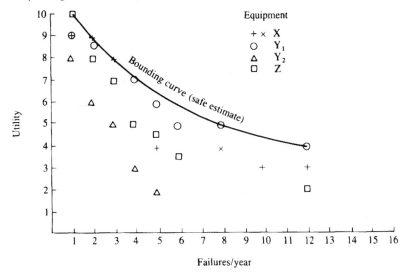

Fig. 5.6 Utility function: failures/year

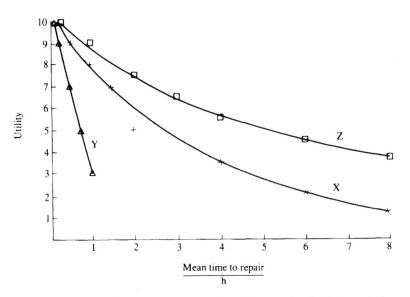

Fig. 5.7 Utility function: mean time to repair. (X, Y, and Z are different applications.)

(ii) reliability (mean time to failure)
(iii) production throughout (kg of product per hour)
(iv) quality of product (percent purity)

All units are similarly priced and the post-purchase support of the suppliers are comparable. Analysis of the design of the units and using

Equipment Evaluation 69

Table 5.1 Performance calculations

Machine	Mean corrective repair time (h)	MTTF (h)	Production rate (kg/h)	Purity (%)
A	1.5	560	1500	90
B	2.0	740	1700	85
C	1.8	620	2000	80

other information from manufacturers has resulted in the estimates given in Table 5.1.

Assume that consultation with experienced personnel has enabled the value expressions, in the form of utility functions as shown in Figs. 5.8(a)–(d).

Using the utility functions, the performances in Table 5.1 can be converted to value scores and the *DPI* calculated using equation (5.5). The results of the calculation are given in Table 5.2, no scale factors were used.

Observations on the DPI *results*
The calculations indicate that Machine B is the best option since it has the highest DPI score. Of course, the decision in reality is not so simple. All the machines look satisfactory since they have good *DPI* scores and, although they are similarly priced, there is no very good reason for not choosing the cheapest.

It is interesting to alter the utility functions, e.g. the slopes of the lines in Figs. 5.8(a) and (b) to investigate if the result changes significantly. In this case it doesn't, but if the calculated result is significantly affected by a small change to the utility function then further investigation into the shape of the function is required to make sure the value judgement is sound.

The method should be used to open up the assessment procedure in order to focus the analyst's attention on important points. Quantitative evaluation methods are not intended to be purely mechanical in operation. They structure the analysis exercise and provide food for thought. The sensitivity of results to changes in the performance calculations and the utility functions should be explored.

Table 5.2 Utility calculations

Machine	Mean corrective repair time	MTTF	Production rate	Purity	DPI
A	7.6	6.6	8.2	9.4	**7.8**
B	6.4	8.3	9.3	8.9	**8.1**
C	6.8	7.1	10.0	7.9	**7.7**

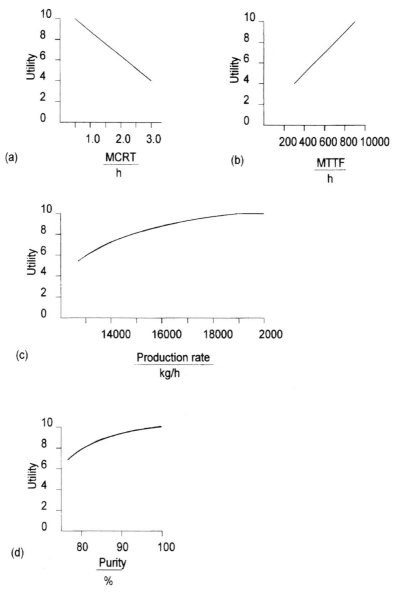

Fig. 5.8 Example utility functions

5.7 Systems analysis using *DPI*

The *DPI* calculation gives an evaluation of the overall usefulness of a piece of equipment as a number in the range 0–10. The calculated answer does not depend on the number of parameters included in the calculation. If an item of equipment is assessed with respect to maintainability, reliability, production rate and production quality and scores, say, 7 with respect to each of the four criteria then the *DPI* calculated is 7. If another piece of equipment is assessed with respect to maintainability, reliability and fuel efficiency and also scores 7 for each of the three criteria, then the *DPI* calculated is also 7.

This leads to the use of *DPI* to analyse and compare items of equipment (functional units) that are each evaluated using different assessment criteria. If all the equipment in a system are all evaluated using *DPI*, then the *DPI* scores will indicate their relative overall performances. *The equipment that has the least score is the least useful to the system.* This approach to system characteristic evaluation is discussed in depth in Chapter 6.

5.8   Equipment handling fluids

5.8.1   *Method*

Particular problems can arise when equipment is used for handling fluids. Corrosion may be a problem. Also, if particulate matter is carried in the fluid even in very small quantities, then there is a possibility of seizure of moving parts if solids are deposited. A useful procedure for the analysis of components handling fluids is:

(i) Identify all mating parts that have relative motion, either during operation or in maintenance operations, then check their sensitivity to seizure.
(ii) Examine the internal contours of the component in the attitude of installation (not simply as drawn) for regions of fluid hold-up. Note if stagnant zones may be formed under conditions of fluid flow.

Flushing with a cleansing agent will not clean a machine if stagnant zones are present. Stagnant zones may also cause problems of corrosion where free-flowing process fluid does not. When a mixture of two or more fluids is present then the potentially harmful effect of one fluid may be inhibited by the action of others mixed with it. However, if a machine is non-draining or if stagnant zones are formed during operation, then there is a possibility of fluid separation occurring. A concentrated layer of liquid may then corrode parts of the component that would otherwise be resistant to the action of the fluid mix. A trap in itself need not necessarily be prohibitive if corrosive liquids are handled, but if the liquid will be held close to seals, screw threads, or highly stressed components, then in-service problems may well ensue.

5.8.2 Brief case study: Ball plug valve analysis

This example refers to the moving parts of a ball plug valve which was proposed for use on a mill used to prepare powders. The mill is a cylindrical drum that contains a mixture of steel balls and powder. The drum is rotated slowly so that the steel balls progressively mill the powder down to the correct grain size. Powder is fed into the drum through a hole in the circumference of the drum which is 'stopped' to seal the drum during the milling process. The drum is 'unstopped' and rotated to empty the powder. The powder is very hazardous and particularly difficult to handle, being very sticky by nature so that it tends to block small diameter pipes. Previous mill designs had used a heavy removable 'stopper' that had caused considerable maintenance difficulties. A new design was proposed which had a conventional process valve, a 100 mm ball plug valve of plastic construction, fixed to the circumference of the drum. A valve of this size was considered to have a good sized bore to allow the powder to flow and the valve requires just a simple 90 degree turn to open and close it to let powder into and out of the drum. It was argued further that the use of a plastic valve would minimize the out of balance force on the ball mill bearings.

In the review of this design proposal, the first point to consider was the moving parts and their sensitivity to seizure since powder flow is involved. The parts that have relative motion during valve operation are:

(i) the valve body and the ball
(ii) the ball and the seals in the valve housing
(iii) the valve stem and the stem packing

Although the life of the ball seats and to a lesser extent that of the stem packing will be reduced by the action of the powder, more important is the sensitivity of the ball to seizure within the valve body. When the valve is partially open, powder will enter the space between the ball and the body and in time the space between the ball and the body will fill with powder. The ball will then only rotate by shearing the powder surrounding the ball circumference. If one assumes that, when the ball rotates, shearing of the powder takes place around the circumference of a complete sphere, then it can be readily shown that the torque required to turn the ball is proportional to R^3 where R is the radius of the ball. Therefore, bearing in mind the sticky nature of the powder being handled, this gives cause for concern about the strength of the plastic valve. The valve size cannot be reduced because the passage through it would be too small. Before proceeding, and on the basis of the above calculation, non-hazardous trials were undertaken using a substitute powder with the required mechanical properties. Repeated opening and closing operations of the valve soon became increasingly difficult as powder entered the space between the ball and the valve body and soon the valve stem fractured. The plastic valve was replaced by one of stainless steel which was strong enough to shear the powder, the stainless steel valve also had the required corrosion resistance. The mill rotated at a slow speed and the calculated radial load on the mill bearings due to the relatively heavy steel valve was within the capacity of the bearings. Although the initial cost of steel valve was over five times that of the plastic valve, a much more costly repair exercise as a result of failure was avoided by undertaking this detailed evaluation.

5.9 Case study: An evaluation of a machine to assemble battery components

5.9.1 Introduction

This case study is concerned with a machine which assembles a set of battery plates prior to their insertion into a battery case. It was decided that a new generation of machines was needed that would have good availability whilst still achieving good production output. The function of the machine is to assemble positive and negative battery plates (approximately $150 \times 150 \times 3$ mm) and a plate separator material into a stack ready for insertion into a battery case.

5.9.2 Assessment criteria

It was first decided that mean corrective repair time, mean time to failure and production rate would be the assessment criteria.

Utility functions for use in the evaluation exercise were then derived for each criterion. This was carried out with reference to the expected factory conditions which were a dirty working environment and a labour force that was not highly skilled. Five members of the company's staff were interviewed independently. They had experience of plant commissioning and operation and of the design and development of similar equipment. For each assessment criterion, each person was asked to give a value score, out of 10, for a range of performances considering the working environment in which the machine would be operated. The extremes of the range of performance from excellent (score $= 10$) to unacceptable (score $= 0$) were first established. Then the intermediate values were quantified by working alternatively from each boundary. Any discontinuity at the meeting at mid-range was smoothed by random questioning throughout the centre range.

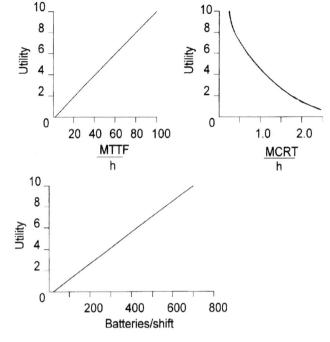

Fig. 5.9   Utility functions for a battery-making machine

5.9.3   *Value judgements*

Utility functions
The utility functions for mean corrective repair time, mean time to failure and production rate are given in Fig. 5.9.

Scale factors
The company personnel wished to use scale factors to rank the assessment criteria. They were obtained using judgements based primarily on experience with similar equipment. Feedback from plant operations had tended to be very critical of reliability rather than production rate therefore mean time to failure was scaled above production rate. Mean corrective maintenance time was not considered as significant as production rate. The value judgements made were reflected in the scale factors:

  Mean time to failure: 3

  Mean corrective maintenance time: 1

  Production rate: 2

5.9.4 Machine analysis

A detailed analysis of the machine design was undertaken using assembly drawings. The study was facilitated by the existence of certain sub-assemblies that had been made up in prototype form as part of an integrated design and development programme.

The nominal failure rates λ_0 were estimated using data in the public domain, reference (**3**), and the use of environment and stress factors to adjust the nominal failure rate to take account of operating conditions (see Chapter 3, Section 3.7). An environmental factor $k_1 = 2$ was applied to take account of the dirty conditions in the factory. The stress factor k_2 will be discussed later. When failure rate data were not available for a particular component an estimate was made after considering the failure rates of similar components. There was no redundancy built into the machine therefore all components were in series.

The mean corrective repair times of components were estimated by reference to general assembly drawings and some prototype sub-assemblies. An experienced technician assisted with these assessments.

Table 5.3 lists the components that were present in the machine. In total 163 components were incorporated into the study.

Using the component count method, see Chapter 3, Section 3.8, the mean time to failure and the mean corrective repair times were calculated.

Mean corrective repair time

The mean corrective repair time was calculated to be 1.32 h working in good conditions. Under the envisaged plant conditions of a dusty atmosphere and a generally dirty environment, these repair times will be exceeded. It was estimated that repairs could take up to four times as long, giving a predicted mean corrective repair time of approximately 5.3 h.

Mean time to failure

Using the component count method, the basic mean failure rate was found to be

$$\lambda_0 = 2.12 \times 10^{-3} \text{ h}^{-1}$$

Applying the environment factor $k_1 = 2$ and using a stress factor $k_2 = 1$ gives $\lambda = 4.24 \times 10^{-3} \text{ h}^{-1}$.

Hence $$\text{MTTF} = \frac{1}{4.24 \times 10^{-3}} = 236 \text{ h}$$

The machine was designed with a variable speed device which could be used to control the production rate. The stress level in the machine

Table 5.3 Components, failure rates, and repair times

Primary drive unit		No. off n	λ (per 10^6 h)	Corrective repair time M_i (h)
(a)	DC Motor (0.5 hp, 2000/73 revs/min)	1	10	0.5
(b)	Spur gear (steel)	2	10	2.0
(c)	Torque limiter	1	3	1.0
(d)	Flexible coupling (polymeric material)	1	10	1.5
(e)	Bevel gear box	2	65	0.75
(f)	Microswitch (lever-cam operated)	6	0.5	1.0
(g)	Keyed rigid coupling	1	0.4	1.5
Shuttle drives				
(a)	Rolling element bearing	2	15	0.5
(b)	Pivot	12	1	0.5
(c)	Adjustment screw	3	0.02	0.5
Separator feed drive				
(a)	Toothed belt	2	40	0.1
(b)	Rolling element bearing	2	15	0.5
(c)	Timing mechanism (light mechanical assembly, a bought-out item with no details provided)	1	11	1.5
(d)	Belt pulley (keyed on)	3	0.2	0.5
(e)	Knurled roller: (i) rolling element brg.	2	15	2.0
	(ii) coil spring	4	1	2.0
	(iii) knife	1	50	2.0
(f)	Idler roller: rolling element brg.	2	15	1.5
Shuttle carriages				
(a)	Linear bearings	8	20	2.5
(b)	Torsion spring (for pawls)	2	0.2	0.75
(c)	Self-lubricated bushes (plate grasp)	2	5	1.0
(d)	Compression springs: (i) plate grasp	1	1	1.0
	(ii) plate adjust	1	0.2	0.5
Plate feeders (proprietary design)				
(a)	Air cylinders	10	3	1.0
(b)	Lubricated bush	26	5	1.0
(c)	Pivot	2	1	1.0
(d)	Pin/reamed hole	18	15	1.0
(e)	(i) Brass bushes (plate stop, one side only)	2	5	1.0
	(ii) Tension spring	1	0.2	1.0

Stacked plates conveyor

(a)	Motor gear unit (0.25 kw, 1400 revs/min motor)	1	10	0.3
(b)	Chain	1	1	0.25
(c)	Chain wheel	3	0.2	0.25
(d)	Rolling element bearing	4	15	0.75
(e)	Conveyor belt (polymer, endless)	1	200	1.5

Pneumatic controls

(a)	Solenoid operator 'on–off' valves: solenoid / valve	6	3	0.5 / 0.5
(b)	Controller	1	300	1.0
(c)	Junction box	1	5	0.5

Stacked plates handling

(a)	Chain	2	1	2.5
(b)	Chain wheel (keyed to shaft)	4	0.2	2.5
(c)	Toothed belt	2	40	2.5
(d)	Belt pulley (keyed on)	4	0.2	2.5
(e)	Rolling element bearings	8	15	2.5
(f)	Width adjustment screw	1	0.02	2.5
(g)	Stepping motor drive	1	5	2.5
(h)	Spur gear: (i) nylon	1	100	2.5
	(ii) steel	1	10	2.5

components depends upon how fast the machine is operated therefore the stress factor k_2 and hence mean time to failure is a function of production rate. It was decided that a speed setting that would give 400 batteries/shift would not be overloading the machine and this production rate was set as the nominal rating. Tests using the prototype sub-assemblies confirmed this decision. Therefore the mean time to failure can be calculated as a function of production rate using different stress factors k_2 for higher production rates, see Table 5.4.

5.9.5 Device performance index

The calculated values of mean corrective repair time, time mean time to failure and production rate were converted to value scores and the *DPI* calculated as a function of production output using equation (5.6). The results are given in Table 5.5 from which it is possible to decide the

Table 5.4 MTTF as a function of machine output (batteries/shift)

Batteries per shift	k_2	MTTF (h)
640	8	29
560	4	59
480	2	118
400	1	236

Table 5.5 Device performance index as a function of machine output

Batteries/shift	DPI
640	4.1
560	6.3
480	7.8
400	7.2

optimum running speed of the machine. It can be seen in Table 5.5 that the *DPI* is maximum at a production rate of 480 batteries/shift.

5.9.6 *Decisions*

It was decided that a variable speed device would not be fitted to production machines and they would be supplied at the optimum fixed production rate setting of 480 batteries/shift. This was done to prevent on site adjustments at a later date to give higher production rates that would reduce reliability and perpetuate the poor reliability reputation of this class of machine.

The predicted mean corrective repair time was 5.3 h under plant conditions giving a utility score of 6.3. This value is rather low and draws attention to the difference between 'what the engineers wanted' and 'what was actually produced'. Several aspects of detail design were identified where improvements to maintainability could be made. Typically, poor maintainability resulted from lack of attention to detail design causing accessibility problems especially when sub-assemblies were mated after being designed independently. Certain features were redesigned for the production machines.

References

(1) Pugh, S. (1991) *Total design* (Addison-Wesley).
(2) Isaksen, S.G., Dorval, K.J., and **Treffinger, D.J.** (1994) *Toolbox for creative problem solving*, Creative Problem Solving Group, USA.
(3) Green, A.E. and **Bourne, A.J.** (1972) *Reliability technology* (John Wiley).

Chapter 6

System Evaluation: Parameter Profile Analysis

6.1 Introduction

Parameter profile analysis is a method of system evaluation that brings out the character of the system with respect to reliability. Maintainability is also included in the method. It may be used to evaluate manufacturing and process plants or to evaluate an item of equipment that itself can be considered to be a 'mini system'. In a design review exercise, parameter profile analysis is an appropriate method of analysis once items have been designed (or selected). It may also be used to evaluate existing plant when considering plant enhancement or a change of conditions.

Research has indicated that large, expensive systems that comprise many relatively low cost items may often be subjected to quite superficial design reviews (**1**). This may well be because the items that form such systems are usually relatively well known, e.g. pump and valve sets, and consequently are not examined in close detail. In contrast, systems that comprise a few relatively high cost items receive much attention in design reviews, probably because high expenditure on individual items attracts attention. The evaluation method given here is suited for use on systems that comprise many relatively low cost items.

The aim of the evaluation method is to identify weak spots in a system and to highlight those areas where the system performance is near a limit. Thus, it identifies potential causes of poor reliability.

6.2 Performance parameters

For each item of equipment in a system there will be a set of performance parameters that define the overall performance of the equipment, e.g. temperature range, pressure rating, output, flow rate. Some parameters, such as pressure and temperature, may be common for different items whereas other parameters may apply only to one item, e.g. a filtration efficiency. The performance parameters that define a system may be

	Plant items					
Parameters	x_{11}	x_{12}	x_{13}	x_{14}	$\cdots$	x_{1j}
	x_{21}	x_{22}				
	x_{31}	x_{32}	x_{33}			
	x_{i1}	x_{i2}	x_{i3}	x_{i4}	$\cdots$	x_{ij}

Fig. 6.1   Parameter profile matrix

described in matrix form with respect to the items of equipment as shown in Fig. 6.1. Note that the matrix will not be full.

When an operating performance requirement moves beyond the performance limit of an item of equipment, e.g. an operating pressure exceeds the pressure limit of a valve, then the system fails. Therefore the proximity of the required operating performance to the limit of capability of the item of equipment is important. The proximity of required performance to performance capability relates closely to a measure of the safety margin. In the case of a system enhancement exercise, the proximity to a limit may indicate an inhibitor on proposed changes.

A data point x_{ij} in Fig. 6.1 that refers to the proximity to a performance limit may be obtained as follows. For an item, the limits of its capability C_{max} and C_{min} with respect to each performance parameter are first determined. Then, the nominal operating point or range at which the item is required to operate is specified. The limits may be represented diagrammatically as shown in Fig. 6.2(a), an example of a valve operating temperature is given. A scale of 0–20 is set against the range of item capability.

The data point x_{ij} that is entered into the parameter profile matrix for this valve with respect to temperature is the minimum value of A and B, that is, the closest the nominal design condition approaches a limit. In this case the data point $A = 4.2$. The value of x_{ij} always lies in the range 0–10. Ideally, when the design condition is a single point at the mid range then the data point is 10. If there is one operating limit only, then the data point is obtained as shown in Fig. 6.2(b). The example given is that of a power requirement for a motorized valve.

Therefore a set of data points can be obtained for each item with respect to the performance parameters that are relevant to that item. Data may be collected from different sources, even from different teams in a large project, and compiled.

6.3 Maintainability

Maintainability applies to all items of equipment and should be included in the matrix of data points. Maintainability performance can be

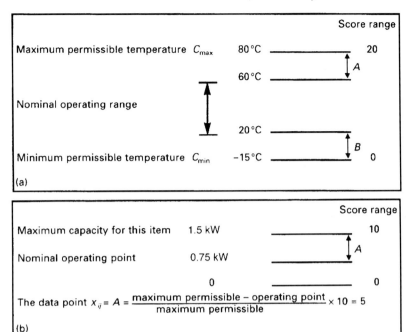

Fig. 6.2 Determination of a data point (a) two limits, (b) one limit

measured by estimating the mean corrective repair time (MCRT) of an item (see Chapter 3, Section 3.3).

In the case of MCRT there are no operating limits that define the performance capability of each item. Instead there is a single prediction of performance of the equipment (an MCRT calculation). Also, there is no MCRT performance requirement for the system to operate, but an expectation of performance with respect to MCRT can be made. The expectation will involve a subjective value judgement.

A data point for maintainability may be obtained as follows. A judgement is made of the repair time M_{min} that is just perfectly acceptable and of the repair time M_{max} which just marks the limit of unacceptability. If the predicted MCRT is M, then the maintainability data point is calculated as

$$x_{ij} = \frac{(M_{max} - M)10}{(M_{max} - M_{min})}$$

For $M < M_{min}$ $x_{ij} = 10$

For $M > M_{max}$ $x_{ij} = 0$.

6.4 Analysis of the parameter profile matrix

Therefore, using the methods described in Sections 6.2 and 6.3 above, a matrix can be compiled containing data points for items relating to all salient performance parameters and maintainability. The matrix may then be analysed by rows and columns in order to evaluate the characteristics of the system.

Consider a row of the matrix. Each data point x_{ij} refers to a single parameter.

(i) Looking along a row reveals whether the system is consistently good with respect to this parameter or whether there is variable performance.
(ii) For a large number of items, a good system should have a high mean and low standard deviation of x_{ij} scores for each parameter.

A parameter performance index (*PPI*) can be calculated:

$$PPI = n\left(\sum_{j=1}^{n} \frac{1}{x_{ij}}\right)^{-1}$$

where $n = $ the number of items in parameter row i.

As discussed in Chapter 5, Section 5.6.5, the inverse method highlights low scores, whereas the addition of scores may hide one if there are some high scores.

The numerical value of PPI lies in the range 0–10 no matter how many data points are included in the calculation of PPI. Therefore, a comparison of PPIs can be made for the system to judge whether the overall performance is acceptable with respect to parameters or whether the system is weak in any respect.

Similarly, for each column, a device performance index *DPI* may be calculated as:

$$DPI = m\left(\sum_{i=1}^{m} \frac{1}{x_{ij}}\right)^{-1}$$

where $m = $ the number of parameters relating to an item in column j.

A comparison of *DPI*s reveals those items that are contributing less to the overall effectiveness of the system. For an individual item, a good result is a high *DPI* and high mean value of x_{ij} scores with a low standard deviation of x_{ij} scores for that item. Of course, individual low x_{ij} scores are readily identified and reveal poor device performance with respect to particular parameters.

System Evaluation 85

6.5 Application of the method: case study

6.5.1 Plant items and raw data

This case study refers to part of a process plant which stores and transfers powdered product between hoppers and bag filling facilities. The system comprised the following items: diverter valve, vent chute, rotary valve, valve vent filter, pressure relief valve, reverse jet vent, blower, non-return valve, bypass valve and silencer/filter (item numbers 1–10 respectively). The system was analysed using design data only. On completion of the design review, the findings were then compared to experience on the plant to check the effectiveness of the method at finding weak system points.

The data points were derived as described above, the two examples Fig. 6.2(a) and (b) above refer to this case study. The items and raw data are given in Fig. 6.3.

6.5.2 Data analysis

The parameter profile matrix was derived from the raw data in Fig. 6.3 and is shown in Fig. 6.4. The matrix is analysed by row and by column. For each row (system parameters), the performance parameter index (PPI) is calculated and for each column (items of equipment) the device parameter index (*DPI*) is calculated. The calculated results for PPI and *DPI* are given in Figs. 6.5 and 6.6. (For this case study there are insufficient data points to justify a statistical analysis.)

6.5.3 Observations on the results

It can be seen that the items are generally satisfactory with respect to maintainability. However, there is a little concern about temperature. The PPI for this variable is 4.2 which indicates that items do not have a lot of 'spare capacity' in this respect. Flow rate is assessed for only one item but item 7 does not have much spare capacity. There is no concern over particle filtration. The main concern is the power variable for which the PPI is 1.6.

Item 6 scores badly with a *DPI* score of only 1.5. The *DPI* is clearly influenced by the low power score. Item 7 also has a poor *DPI* of only 2.7, again it is influenced largely by the power score. The system analysed is vulnerable with respect to power, items 6 and 7 are operating near their limits and might be expected to be trouble spots. These items are, respectively, a reverse jet vent (operating point 2 kW and maximum capacity 2.2 kW) and a blower (operating point 18 kW and maximum capacity 22 kW). Temperature problems may be encountered should there be a need to increase temperature, say if new materials were to be processed.

1. Diverter valve.
   MCRT = 1.5 h
   Temperature: nominal working range = 20 to 60°C. Limit range = −15 to 80°C
2. Discharge vent chute.
   MCRT = 5 h
   Temperature: normal working range = 20 to 60°C. Limit range = −20 to 70°C.
3. Rotary valve.
   MCRT = 4 h
   Temperature: nominal working range = 20 to 60°C. Limit range = −30 to 178°C.
   Power: nominal working point = 0.75 kW. Capacity = 1.5 kW.
4. Rotary valve vent filter.
   MCRT = 8 h
   Particle size: nominal working point = approx 40 μm. Limit = 10 μm.
5. Pressure relief valve (2 off).
   MCRT = 4 h
   Temperature: nominal working range = 20 to 60°C. Limit range = −40 to 100°C.
6. Reverse jet vent filter.
   MCRT = 8 h
   Power: nominal working point = 2 kW. Capacity = 2.2 kW.
7. Blower
   MCRT = 8 h
   Flow rate: nominal working point = 0.26 m^3/sec. Capability = 0.37 m^3/sec.
   Power: nominal working point = 18 kW. Capacity = 22 kW.
8. Non-return valve.
   MCRT = 4 hr
   Temperature: nominal working range = 20 to 60°C. Limit range = −10 to 120°C.
9. By-pass valve.
   MCRT = 4 h
   Temperature: nominal working range = 20 to 60°C. Limit range = −10 to 120°C.
10. Silencer/filter (2 off).
    MCRT = 8 h
    Particle size: nominal working point = approx 40 μm. Limit = 10 μm.

Maintainability
   M_{max} = 13 h
   M_{min} = 1.5 h

Fig. 6.3   Item data

	\multicolumn{10}{c}{Items}									
	1	2	3	4	5 (×2)	6	7	8	9	10 (×2)
MCRT	10.0	7.0	7.8	4.3	7.8	4.3	4.3	7.8	7.8	4.3
Temp	4.2	2.2	4.8	—	5.7	—	—	4.6	4.6	—
Power	—	—	5.0	—	—	0.9	1.8	—	—	—
Flow rate	—	—	—	—	—	—	3.0	—	—	—
Particle size	—	—	—	7.5	—	—	—	—	—	7.5

Fig. 6.4   Parameter profile matrix: Powder discharge facility

	PPI
MCRT	5.9
Temp	4.2
Power	1.6
Flow rate	3.0
Particle size	7.5

Fig. 6.5 Analysis of system parameters

	Items									
	1	2	3	4	5	6	7	8	9	10
DPI	5.9	3.3	5.6	5.5	6.6	1.5	2.7	5.8	5.8	5.5

Fig. 6.6 Analysis of items of equipment

Discussion with plant personnel revealed that the reverse jet vent and the blower had experienced the problems forecast. The power capacity of the reverse jet vent and the blower should be increased. The plant has a good reliability record in all other respects and has a good maintainability reputation. No temperature problems have yet been encountered. In all other respects the plant has a good operating record.

6.6 Discussion

Parameter profile analysis is intended for use when details of the equipment are known. In particular, it is intended for use in schematic design when systems are synthesized using proprietary equipment. Many manufacturing and process plants are designed in this way.

Parameter profile analysis provides a quantitative evaluation that adds value to a design study by integrating performance parameters such as pressure and temperature with maintainability. The performance data required for the evaluation are the same as that used to specify the equipment. Therefore there is no requirement made for extra data in excess of that which would be generated in a good design exercise. The only additional requirement is a value judgement of perfectly acceptable and unacceptable levels of mean corrective maintenance times.

The data are compiled into a parameter profile matrix using data points derived from the proximity of a required operating point to the performance limit of the equipment. The use of the proximity of the performance requirement to a limit of equipment capability relates to the concept of a safety margin. Having compiled the matrix, then analyses can be performed with respect to particular variables for all items of equipment or for particular items of equipment for all relevant variables, yielding the *PPI* and *DPI* indices respectively.

It is a straightforward task to input and analyse the data on proprietary spreadsheet software. On a large project, data can be generated by different design teams and compiled and analysed using the software in the design review exercise. Importantly, the analyst has flexibility in the evaluation and sensitivity analyses may be carried out by varying system performance requirements.

Software flexibility and the method of data point derivation are especially useful in plant enhancement exercises. There are instances when existing plants are subject to changes, for example, there may be advantages in processing a product at different pressures and/or temperatures. The equipment performance data are already available in the software. The analyst may explore implications of changes to temperature and pressure requirement since the parameter profile analysis method will immediately identify when the performance of an item is in close proximity to a limit.

Also, plants may be required to process materials that are different to those which they were originally designed to process. If a parameter profile analysis has been carried out when the system was first designed then the equipment data are already available; all that is needed is to input new process data and the analysis method will identify weak areas. Plant enhancement and plant flexibility are important aspects of the optimum use of process plant assets. The additional time taken to set up the equipment database during a design review will be justified since the data can be used throughout the life of the plant.

6.7 Summary

The parameter profile analysis method can be used effectively to collate and analyse system data to reveal the weak areas of design proposals where problems may be found in service.

The method uses equipment performance data that are based on the proximity of the actual performance required to the performance capability of the equipment. Maintainability is included as mean corrective repair time calculations related to expectations of equipment maintainability.

The method calculates:

(i) overall system performances with respect to particular parameters, and
(ii) overall equipment performances with respect to all relevant parameters

Parameter profile analysis evaluates systems. The principles on which it is based, i.e. the proximity of a performance requirement to a performance limit capability, can also be used to design equipment for high reliability. This is discussed in Chapter 11.

The advantage of using a formal analysis procedure such as parameter profile analysis is to ensure that particular weak points are found. In the brief case study in Section 6.5, the method did ensure that the right questions were asked and weak points were identified. In the original design exercise the lack of a systematic analysis procedure meant that weak points were less likely to be identified before the production started.

References

(1) **McKinney, M.** and **Thompson, G.** (1989) A survey of process plant maintainability problems, *Proc. Instn Mech. Engrs, Part E, J. Process Mech. Engng*, **203**, 29–35.

Bibliography

(1) **Moss, T.R.** and **Strutt, J.E.** (1993) Data sources for reliability analysis, *Proc. Instn Mech. Engrs, Part E, J. Process Mech. Engng*, **207**, 13–19.
(2) **Thompson, G., Goemiine, J.,** and **Williams, J.R.R.** (1998) A method of plant design evaluation including maintainability and reliability, *Proc. Instn Mech. Engng, Part E, J. Process Mech. Engng*, **212**, 71–80.

Chapter 7

Failure Mode Analysis

7.1 Introduction

Failure mode analysis is an important part of design. Consideration of how a machine or system *may* fail is a first step to achieving good maintainability and reliability. In general, components are designed not to fail within the expected life of equipment. Some designers are reluctant to consider component failure since they use their talents to prevent failure. However, failure mode analysis is not an admission of inferior design ability. Rather, it is a recognition that failures do occur, for whatever reason, and that it is a good idea to 'step back' from the design to consider the possibilities of what might go wrong.

Failure mode analyses may be undertaken with different degrees of complexity. A simple approach can be used in concept and detail design that is not time consuming but which will yield high dividends with respect to maintainability and reliability. There are also comprehensive, detailed methods that are appropriate for use in safety assessment exercises. Both approaches are useful provided that they are used appropriately. It would be quite wrong to embark on a time consuming, comprehensive analysis every time failure modes are considered.

A simple approach to failure mode analysis with respect to maintenance and reliability will be discussed first. More comprehensive approaches, which are used in safety assessments, will then be described.

7.2 Failure mode and maintenance analysis (FMMA)

7.2.1 Outline method

In this method, failure mode analysis is used to identify the principal areas for which the required maintenance and repair actions should be especially considered and evaluated.

The principal failure modes of the equipment (or system) are first identified. Not every component needs to be considered, except for the simplest machines that comprise very few components. What is required are the main areas, perhaps numbering five or six, that are most likely to cause problems in service. Judgement based on experience is beneficial.

Then, for each principal failure mode:

(i) The actions and procedures that are required to repair the equipment are studied to ensure that they are as straightforward and as simple as is reasonably practicable.

(ii) The feasibility of any routine maintenance procedures or condition monitoring methods that may be required, even at a later stage, are thought through and due allowance made for them in the design.

Of course, in addition, particular attention should be paid to the strength of components to prevent failure.

FMMA can be used to evaluate design proposals or proprietary items of equipment.

7.2.2 FMMA in concept design

The objective of FMMA in concept design is to identify those areas where particular attention should be paid to achieve high maintainability and reliability. A the concept stage there is insufficient detailed information available to study specific maintenance actions.

For example, consider the mechanism shown in Fig. 7.1. A pneumatic actuator is connected to the mechanism at C and displaces link A–B through an angle. The four bar mechanism then controls the motion of point D which slides on a linear bearing. At point D, a bag is placed which is moved between positions X and Y. The empty bag is placed on D at position Y, the bag is moved to position X to fill it and then moved to Y for removal from the machine.

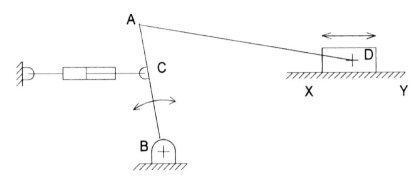

Fig. 7.1 Four bar mechanism

The principal failure modes are:

(i) linear bearing failure (high wear is expected due to powder spillage on the bearing and slide at X)
(ii) pneumatic actuator failure (main actuating mechanism with seals and moving parts), and
(iii) pin joint failure at C (high load point on link A–B)

Therefore, special attention can be given to these components to size and protect them correctly to avoid failure and to ensure that they can be repaired easily should failure occur. For example, the joint C could be made as an integral part of the pneumatic actuator incorporating a quick release feature from link AB.

7.2.3 *FMMA in detail design*
In detailed design analysis, FMMA can be used to study particular maintenance and repair actions. A useful approach is to consider the principal elements of a design proposal (or proprietary machine) rather than every detail which would be prohibitively time consuming in many cases.

Consider the centrifuge design shown in Fig. 10.15. The basic elements of the design can be expressed simply as shown in Fig. 7.2.

The principal failure modes are:

(i) bearing failure
 – tight bearings
 – worn bearings
(ii) bowl attachment failure
 – detached bowl
 – loose bowl
(iii) break in the magnetic coupling drive

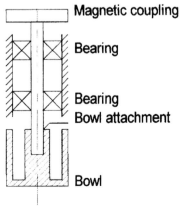

Fig. 7.2 Centrifuge design

For each failure mode, the required repair actions for these failure modes can then be assessed to determine if they are acceptable. The repair actions are discussed in Chapter 10, Section 10.5.7 and so will not be discussed further at this point.

7.2.4 FMMA and condition monitoring

FMMA can be used as a basis to select appropriate condition monitoring techniques.

The method is as follows:

(i) Determine the principal failure modes.
(ii) Derive a morphological chart (see Appendix 2) that shows the alternative methods by which each failure mode may be detected.
(iii) Choose an optimum set of monitoring methods that enables the failure modes to be detected using the minimum number of monitoring methods.

Example

For the centrifuge concept described in Fig. 7.2, the morphological chart of condition monitoring methods is as shown in Table 7.1.

The condition monitoring methods were selected for the centrifuge on the following basis:

(i) Each failure mode should be detected by at least two methods.
(ii) A minimum number of monitoring methods should be used.

The basis of the selection can, of course, be changed according to circumstances. The important points are the use of failure mode analysis as the starting point and the use of a morphological chart in a creative approach to the selection of the most appropriate condition monitoring methods.

In the case of the centrifuge, the following condition monitoring methods were selected.

(i) An accelerometer mounted on the bearing housing to detect high vibration levels caused by a loose bowl or worn bearings.
(ii) A capacitance transducer to detect increased shaft displacement during operation caused by worn bearings or bowl attachment problems.
(iii) A temperature transducer adjacent to the lower bearing to detect a tight bearing. A high temperature in this bearing may cause lubricant to leak and contaminate the process fluid. Also, it is the higher loaded of the two bearings.

Table 7.1 Morphological chart showing centrifuge failure modes and condition monitoring methods

Fault	Condition monitoring methods			
Tight bearings	Increase in motor current	Lubricant leak detection	Transducer to monitor the rise in outer race temperature	Thermal image of complete machine
Worn bearings	Vibration monitoring of bearing housing	Lubricant leak detection	Increased displacement of shaft	Acoustic emission
Detached bowl	Vibration decrease	Process fluid quality (no separation)	Process fluid quality (presence of debris)	
Loose bowl	Vibration increase	Process fluid quality (presence of debris)		
Break in the magnetic coupling drive	Decrease in motor current	Change in shaft vibration characteristic		

(iv) Motor current monitoring to detect a detached bowl, coupling drive failure or tight bearings.
(v) The product quality is monitored routinely and would detect a detached bowl because solids would not be separated from the liquid.

Three instruments are used and the motor current and product quality are monitored. Each failure mode can be detected by at least two methods. The methods seek to achieve a high reliability of detecting a fault rather than to pin-point a particular cause of failure during operation. If a fault is detected, then the machine would be stopped and examined to determine the cause of the problem.

7.3 Risk and risk assessment

The risk associated with an event is the product of the frequency (or probability) of the event occurring and the consequence that results if the event occurs. Risk is defined as:

$$\text{Risk} = (\text{the frequency of an event}) \times (\text{the consequence of the event}) \tag{7.1}$$

Two events that are quite different in character may therefore have the same risk using equation (7.1). An event that occurs infrequently but with high consequence can have the same risk as an event that occurs more frequently but with lower consequence.

In the evaluation of machinery and systems, there will be many events that can occur such that each event gives rise to a particular consequence. In these cases, the total risk for n events is:

$$\text{Risk} = \Sigma_i^n \, (\text{frequency})_i \times (\text{consequence})_i \tag{7.2}$$

Risk is commonly associated with harm to human life or the environment, but not exclusively so. One may consider the risk of causing downtime in a production operation.

Considerations of risk pose difficult questions for designers. What is acceptable? Two events that have the same risk according to equation (7.1) may not be equally acceptable, even though they may have the same calculated risk. For example, an event that results in high downtime, even if reliability calculations show that such an event is an infrequent occurrence, may not be acceptable, whereas a more frequently occurring event with low downtime may be acceptable.

One method which the severity of a consequence can be taken into account is by modifying equation (7.1) as follows.

$$\text{Risk} = (\text{the frequency of an event}) \times (\text{the consequence of the event})^c \quad (7.3)$$

The power c increases the significance of the consequence of an event. The value of c depends upon the subjective view of the analyst but it should not be made too great and would be expected to lie in the range 1.0 to 1.5. The modification can be applied similarly to equation (7.2).

When considering risk in design, the emphasis should be equally on limiting consequence and reducing frequency, the latter by improving reliability. One need not carry out quantitative calculations every time. If the designer is aware of risk as a function of *probability* and *consequence* during design practice, just as one is aware of stress combinations, stress concentration factors etc. without performing calculations in every case, then better designs will be produced.

One can conceive of an acceptance criterion for assessing the risk of an event that is a function of the frequency of an event and the consequence, such as the linear equation:

$$\frac{\text{frequency}}{a} + \frac{\text{consequence}}{b} \leqslant 1$$

where a = the maximum frequency that can be tolerated even if the consequence is, for all practical purposes, negligible.

b = the maximum consequence that can be tolerated even if the frequency is, for all practical purposes, negligible.

If an event has $\dfrac{\text{frequency}}{a} + \dfrac{\text{consequence}}{b} > 1$ then it is unacceptable.

The concept is illustrated in Fig. 7.3. Individual events can be reviewed with respect to the frequency – consequence acceptance criterion to determine if a design proposal contains an unacceptable risk before assessing the overall risk using equation (7.2).

Safety assessment (safety = 1/risk) is a specialist subject that is beyond the scope of this book. For a study of safety assessment see reference (**1**).

A hazard is an event that may cause harm. Risk assessment, safety assessment and hazard analysis (known sometimes as 'Hazan') are terms that are used to describe the same basic process, that is, the identification of hazards, the probabilities that the events will take place and the consequences that result if the events occur. The use of hazard analysis in chemical plant operations is described in reference (**2**).

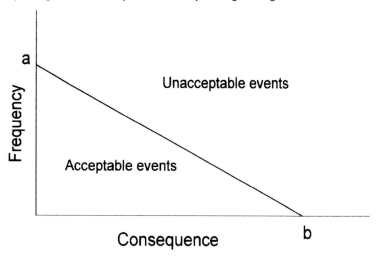

Fig. 7.3 Frequency–consequence acceptance criterion

7.4 Failure mode and effect analysis (FMEA)

FMEA is a widely used analysis method that is based on the two elements of risk: *frequency* and *consequence*. It involves the systematic evaluation of the possible failure modes of a machine or system, their likelihood and the consequences of failure. The failure modes are ranked in order of severity in order to identify the most critical and corrective action is considered for the most severe cases. FMEA is known as failure mode effect and criticality analysis (FMECA) when the analysis is extended to include the criticality of failure modes.

The method begins at the most detailed level that is practicable. It may be applied to systems or machines. In comprehensive equipment studies, an inventory of almost every component may be made.

The outline FMECA procedure is as follows:

(i) Define and list the failure modes at the most detailed level that is practicable, considering all types of failure, e.g. excessive wear, fracture, mal-operation, premature operation etc. For some equipment, the failure modes may be the failures of all the components that make up the equipment.

(ii) Assess the likelihood of failure.

(iii) Study the effect of the failure, considering: harmful consequences, loss of production, secondary effects etc.

(iv) Assess the seriousness of the failure consequences.

(v) Determine the most critical failure modes on the basis of the likelihood of failure and the consequence of failure.

(vi) Decide if action is required to reduce the likelihood or consequence of failure.

Different methods are used to determine the most critical failure modes. One method is:

(i) Score the likelihood of failure of each failure mode on a scale 1–10 (10 = most severe).
(ii) Score the consequence of failure on a scale of 1–10 (10 = most likely).
(iii) Criticality is then calculated as:

$$\textit{score of likelihood of failure} \times \textit{score of consequence failure}$$

(most critical case = 100), which follows the definition of risk as given in equation (7.1).

FMEA can be used for a variety of purposes. It is used as a review of failure modes and their effect on the production availability of equipment. It is an analysis method that complements reliability studies. Whenever there is the possibility of failures, whether the failures be parts of a machine or of a system, then FMEA can be used to evaluate the seriousness of the consequences of the failures, whatever they may be.

Comprehensive FMEA studies are used extensively to evaluate the safety of equipment and systems. In certain cases, where there is sufficient data, mean failure rate data are used to assess the likelihood of failure and, in the most serious cases, the consequences of failure may be expressed in terms of loss of human life. Most safety studies involving FMEA are not so specific, but they do consider a wide variety of causes of failure and their consequences in detail. Therefore they can be extremely time consuming and, although the method is effective, its use cannot be advocated in every case. For further details of FMEA studies applied to safety assessment see reference (**3**).

However, in design evaluation, the outline procedure of FMECA given above can be used effectively as a quick assessment of equipment provided that the study can begin with the principal failure modes rather than at the most detailed level. The group of the most critical failure modes are first identified by engineering judgement and FMECA is then used to rank the list and consider any required actions. Experience is important. The possibility of certain significant component failures being overlooked is reduced since the designer has already considered the loads on components. A safety analyst may well have to consider more detail in order to gain a deeper understanding of the equipment.

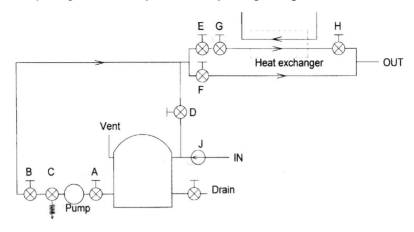

Fig. 7.4 Simplified part of a process plant

Example

Figure 7.4 shows a small part of a simplified process line in which fluid is pumped from a buffer storage tank through a heat exchanger or via a bypass line. Valves A, B, G and H are isolation valves. Valve C is a pressure relief valve, valves D, E and F are flow control valves and valve J is a one way (check) valve. D would normally be closed but it could be opened to redirect flow back to the storage tank.

An FMECA analysis would include the events and analysis shown in Table 7.2. This is a simplified analysis and is just a brief overview of the failure modes, their consequences and probabilities. The determination of the consequence and probability ratings requires much thought and thorough investigation involving: the detailed design of the items of equipment that have been selected or designed; the loading on the equipment; the environmental conditions; the hazardous nature of product that is processed, etc. In this example, the criticality rating identifies the pressure relief valve as the most critical item, as expected, and the mal-operation of the control valves. The pressure relief valve would be subjected to regular in-service testing and the control system should be checked thoroughly prior to installation, monitored for a trial period and subjected to periodic checks. The pump and control valves rank next highest and condition monitoring should be considered for these items.

Even in this simple case, it can be seen that the analysis of failure modes is a complex affair. A moment's consideration of the system will reveal that the possible failure modes and the interaction between elements of the system is far more complex than that given in Table 7.2. For example, the prospect of overfilling the buffer tank has not been

Table 7.2 Failure mode and effect analysis

Component	Failure mode	Consequence	Probability rating (y)	Consequence rating (x)	Criticality rating (x, y)
Valve A	Stuck open	System drained to replace pump	1	2	2
Pump	Mechanical failure	System downtime	3	2	6
Valve B	Stuck open	System drained to replace pump	1	2	2
Pressure relief Valve C	Failure to relieve pressure	System over-pressure	3	10	30
Valve D	Fails to open	(i) Fluid leak from system downstream	1	(i) 2	(i) 2
		(ii) Possible over-pressure		(ii) 10	(ii) 10
Control valve E	(i) Fails to move accurately	(i) Loss of process fluid temp control, wasted product	(i) 2	(i) 7	(i) 14
	(ii) Fails closed	(ii) In conjunction with valves F, D, and C, an over-pressure situation possible	(ii) 1	(ii) 10	(ii) 10

(continued)

Table 7.2 (continued)

Component	Failure mode	Consequence	Probability rating (y)	Consequence rating (x)	Criticality rating (x, y)
Control valve F	(i) Fails to move accurately	(i) Loss of process fluid temp control, wasted product	(i) 2	(i) 7	(i) 14
	(ii) Fails closed	(ii) In conjunction with valves F, D, and C, an over-pressure situation possible	(ii) 1	(ii) 10	(ii) 10
Valve G	Stuck open	System drained to replace heat exchanger if required	1	2	2
Valve H	Stuck open	System drained to replace pump	1	2	2
Heat exchanger	Internal leakage	(i) Contamination of process fluid by water	(i) 1	(i) 7	(i) 7
		(ii) Contamination of water by process fluid	(ii) 1	(ii) 8	(ii) 8

Table 7.2 (continued)

Component	Failure mode	Consequence	Probability rating (y)	Consequence rating (x)	Criticality rating (x, y)
Pipe joints	External leakage	Environmental contamination	1	8	8
Valves E and F	Closure of valves E and F through mal-operation	Over-pressure incident caused	3	10	30

considered nor has the failure of the tank vent. Both events would have very serious consequences. Therefore the use of FMEA is time consuming if it begins at a detailed level.

7.5 Fault tree analysis (FTA)

FTA is a 'top down' approach that begins with certain system failure modes, e.g. failure to deploy the landing gear of an aircraft, and then asks the question 'What sequence or combination of failure events must happen to cause this failure mode?' The sequence of failures are structured so that the relationship between the events can be clearly understood. FTA can be used to identify critical areas in a design review. (In contrast, FMEA is a 'bottom up' approach that begins with an itemized list of components and asks the question 'What are the consequences if this component fails?')

Figure 7.5 shows a simple mechanism in which the motion of link A–B is controlled by a hydraulically actuated cylinder. The hydraulic pressure is provided by either an electrically operated pump or a hand pump. If one considers the system failure 'Failure to return link A–B', then the event structure shown in Fig. 7.6 can be identified. Note the use of 'OR' and 'AND' gates.

A quantitative evaluation of fault trees may be undertaken. By using the failure probabilities of each failure event and the logical relationships between events, the probability of a particular system failure mode may

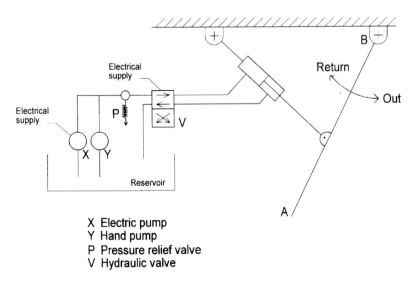

Fig. 7.5 Hydraulically actuated mechanism

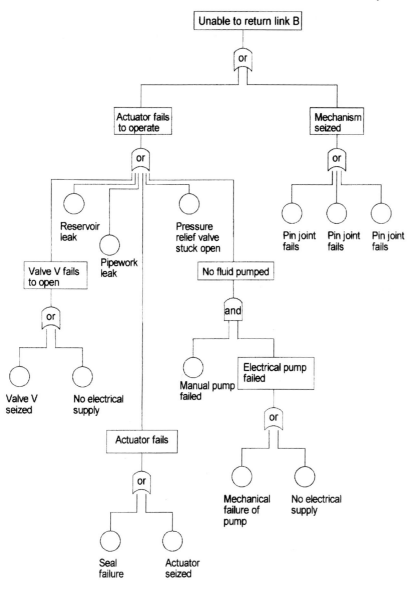

Fig. 7.6 Fault tree analysis

be found. For further information on quantitative assessment and on the notation used to formulate fault trees see reference (**3**). The quantitative analysis of systems by FTA is a detailed analysis method that requires a large amount of design work to be completed to provide sufficient data for analysis. It is therefore a checking procedure to validate designs rather than a method to be used in creative design work.

It can be seen that, even for the simple system shown in Fig. 7.5, the fault tree can become quite complex. The complex nature of FTA is an obstacle to its wider adoption and computer methods have been proposed that generate fault trees for complex systems, see reference (**4**).

7.6 Hazard and operability (HAZOP)

A HAZOP study is one that identifies the hazards that may be present, whereas a risk assessment considers the likelihood and consequences of hazards. A HAZOP includes faults in systems and, importantly, the mal-operation of equipment and systems. It is a technique that is carried out post-design and is a very useful method that enables plant personnel to understand what can go wrong with systems. HAZOP studies may be undertaken at different stages in design. Early on in the design process, systems may be analysed before functional units have been designed. When equipment has been designed, a comprehensive HAZOP study may be undertaken.

In a HAZOP study, a particular set of words are used to stimulate thought to identify hazards. For example, if one were studying the flow in a chemical plant the words NONE, MORE OF, LESS OF etc. would be used. Reference (**2**) gives a comprehensive account of HAZOP procedures for chemical plant analysis.

However, in general, the principles of HAZOP apply to all equipment. It is difficult to be prescriptive about the set of words that would stimulate thought about all types of equipment. But, there is no reason why the designer should not 'step back' from his design at the concept stage and/or the detail design stage and consider what the hazards might be, not just from a failure perspective but from an operability perspective also.

Typically, the following questions are illustrative of what might be in a general HAZOP study.

- What actions are required to start the machine safely?
- What actions are required to shut down the machine safely?
- If a machine is switched on inadvertently, what damage might occur?
- If control valves are present can their incorrect use cause harm?
- Can the closure of a valve or other blockage cause an over-pressure incident?
- What happens to applied loads if the resistance is removed?
- What happens if the machine is overloaded?

- Can out of specification parts be accommodated in manufacturing operations?
- Can harmful materials be ejected from the system?
- What happens if the equipment is operated empty?
- What happens if the equipment is left for a time without operating it?
- What happens if the equipment is operated independently of other equipment?

For particular applications, it should be possible for designers to take an objective view of their work and consider the hazards that may result from both equipment failure and incorrect operation. The latter may be due to operator error or problems caused by manufacturing or process conditions being out of specification.

7.7 Summary

Failure mode analysis has been developed in many different directions and increasingly complex methods are being developed. However, in most creative design work it is awareness of the basic principle of risk that is important rather than the detailed method of a particular analysis procedure.

If, during design work, whether it be in concept design or detail design, consideration is given to the *probability* of an event occurring and its *consequence*, then the quality of design work will be improved.

The use of in-depth FTA and FMEA analyses is the job of the specialist analyst. In-depth studies can only be undertaken after detail design work has been undertaken, for until that stage there is no information to work on. Therefore such detailed methods are checks on the quality of design rather than aids to good design.

Conceptual design is where the character of a design is born, it is where the fundamental problems are sometimes created. For the concept designer, it is the principle of risk assessment that is significant rather than knowledge of a complex analysis method.

For the detail designer, a risk assessment of items, say a mechanism or valve, using FMEA or FTA is a reasonable proposition as an aid to the selection of a component for use in a design scheme. FMEA and FTA studies are useful at the detailed level even without the use of quantitative data. Alternative items may be analysed and compared just by considering the failure modes and their consequences. Either a 'top down' FTA or a 'bottom up' FMEA study will prompt the detail designer to ask the right questions in order to compare the designs.

In conclusion, failure mode analysis and knowledge of the principles of risk are important aids to achieving good maintainability and reliability through design at both concept and detail design stages. However, the use of in-depth analysis methods is a check on completed design work and is usually undertaken by specialist analysts.

References

(1) Thomson, J.R. (1987) *Engineering safety assessment* (Longman Scientific & Technical).

(2) Kletz, T. (1992) *Hazop and Hazan identifying and assessing process industry hazards* (IChemE).

(3) Davidson, J. and **Hunsley, P.** (1994) *The reliability of mechanical systems* (Mechanical Engineering Publications).

(4) Andrews, J. (1994) Optimal safety system design using fault tree analysis, *Proc. Instn Mech. Engrs, Part E, J. Process Mech. Engng*, **208**, 123–131.

Chapter 8

Specifications, Contracts, and Management

8.1 Introduction

At the beginning of any design exercise the designer, or design team, has to follow a set of instructions. Commonly, the instructions are referred to as a 'design specification'. However, the term should not be confused with a comprehensive, detailed, itemized 'specification' that defines the finished article or production system.

A design specification is a set of requirements and constraints that direct the designer. It sets objectives and requirements that must be met and requirements that are desirable. It may contain instructions, points of concern and descriptions that capture the vision or essence of the client's requirements. A good design specification is needed to set the course for a successful design exercise. Sometimes a design specification is known as a 'design brief'.

This chapter is about the description and definition of maintainability and reliability requirements at the beginning of a design exercise. The objective is not to be pedantic about the format of a specification, but rather to discuss how maintainability and reliability criteria may be incorporated into companies' own design specifications.

The design specification often forms the basis of the contract between the designer and the client. Therefore design specifications and contractual matters are often inextricably linked. When disputes occur, the first question asked is usually 'What does the design specification say?' The significance of certain aspects of design specifications on contractual agreements will be discussed where appropriate.

8.2 Design specifications: general principles

When writing a design specification, the aim is to generate a document, agreed between designer and client, which contains:

- a statement of the design objective
- a set of functional requirements
- the constraints on the design

Creative thinking is required. Only by a thorough investigation of the problem can a design specification be derived competently. A long list of general points in a specification is not very helpful to the designer. Very significant aspects of the specification may be lost in the generality. Those aspects of the specification which are essential should be highlighted but they should be kept to a minimum. The objective is to direct the designer but to leave sufficient freedom to generate and develop ideas in order to achieve a good solution.

The specification should specify *functional requirements* and normally particular solutions should not be included. For example, it would be correct to say that an actuator should:

- have a particular speed of operation
- be able to provide a certain force, and
- move with a required positional accuracy

But, it would be wrong to name a particular actuator type, e.g. hydraulic, pneumatic etc. as that might over-constrain the design. Similarly, a specification may require minimum corrosion or wear rates but the designer would normally be left to specify the materials.

However, there are circumstances when the design specification should state particular solutions. For example, if experience has shown that products from a certain manufacturer have proved reliable or that the existing workforce have expertise in maintaining and repairing certain equipment then it makes sense to use these products again. The reasons for doing so should be explained in the specification so that the designer is fully informed. Also, some material choices can be made in the specification, for example, a particular quality of steel may be specified because its corrosion resistance has proved beneficial in similar circumstances. Sometimes a specification may be worded as '...of equal or better performance than X...' which gives the designer scope for creative thought.

8.3 Maintainability and reliability objectives

8.3.1 Meaningful statements

Design specifications should contain meaningful, precise statements of requirements. General comments that refer to 'good quality design work' and requirements that are open to interpretation will achieve little and may even add to difficulties later if the client is unsatisfied with the design work carried out. The author has been engaged in legal disputes concerning 'What is reasonable quality design work' and the outcome is often precarious for all the parties involved.

It is unacceptable to specify that maintainability should be 'as good as possible', 'reliability should be maximized', 'equipment should be maintainable', 'high reliability should be achieved' or other vague remarks. Such statements are of little use to designers. How good is 'good maintainability' or 'high reliability'? In a design specification the requirements should be stated unambiguously if they are to be meaningful. A designer may believe that good reliability has been achieved but the client may disagree. Alternatively, a designer may produce an expensive design to achieve a certain level of reliability but the client may feel that the additional cost is not justified and a lower level of reliability would have been acceptable.

The terms used in a design specification should be precise. For example, if the term 'mean repair time' is used then the meaning of the term should be clarified. If the term it is intended to define 'the mean time to return a machine or system to a normal operating condition', then it must be recognized that the time includes a major element that is not within the designer's control. That is, the time to obtain spares and organize manpower which may be well over 50 percent of the mean repair time. In a contractual dispute over a long mean repair time, the designer may claim that the corrective repair time for which he was responsible is satisfactory and that the maintenance management team should 'sharpen their act' to reduce the time to organize manpower etc.

8.3.2 *Relationships between design objectives*
When a design specification is derived, the relationships between objectives must be considered.

In general, it is not possible to maximize both maintainability and reliability. In equipment design, maintainability can be improved by making components replaceable in the event of wear or damage and by incorporating features such as quick release devices. The additional components such as seals and fasteners etc. that are introduced in this way have the effect of reducing the reliability of the equipment. The objective of the designer is to compensate the drop in reliability by an increase in maintainability so that an overall increase in equipment availability is achieved. Therefore, an objective to maximize maintainability and reliability is not the right approach. It may be that, given certain maintainability features, the required level of reliability has to be achieved by the use of high quality or oversize components and/or by introducing redundancy into the system which could well increase cost.

Availability is attractive as a requirement since it incorporates both maintainability and reliability considerations. It is possible to use availability in a design specification but an estimate is required for the maintenance management times to convert mean corrective repair times to total mean repair times. The additional time estimates may prove unattractive to many. In any case, even if availability is considered, it is probably better to determine the required reliability and mean corrective repair times that designers influence and to write them specifically into the design specification.

Cost, maintainability and reliability requirements should be compatible. The achievement of high reliability levels does not always mean that very high costs will be incurred. However, it is unrealistic to set very low cost requirements and simultaneously expect high levels of reliability. Low cost targets will mean that cheap, inferior components have to be selected which results in adverse effects on reliability.

8.3.3 Understanding requirements

Design requirements that refer to maintainability and reliability should be explicit and not implicit. They should be stated independently of other requirements, where possible, and not be embedded within a list of other requirements where they will not gain the significance they require.

The maintainability and reliability requirements that are made in a design specification should be made in terms that can be readily understood. This is especially important when designs are put out to contract. There are occasions when certain terminology is readily understood in some industries but which may be unfamiliar to design contractors. An example taken from a design specification is a reference to 'unit parts', which the designer contractor failed to understand. No one thought to explain the meaning as it was in common use in the client's company. The outcome was unhappy for all parties as the resulting design was unfit for its purpose.

Environmental conditions should be explained in the specification if they have a particular influence on maintainability and reliability. Reliability calculations may be affected. Maintenance and repair actions may be influenced. Every effort should be made to explain the circumstances that dictate particular requirements for maintainability and reliability in design specifications.

Some contracts require that the contractor must fully inform him/herself of the specification. However, the client should explain and emphasize special requirements. Recourse to legal action following poor design work that results from misunderstandings indicates a poorly managed design project.

8.4 Contents of a specification

8.4.1 Quantitative requirements

The following quantitative requirements should be specified whenever possible.

(i) For reliability, a mean failure rate may be specified for a system or machine. In some cases, a minimum mean failure rate for individual, critical items of equipment may be specified in addition to an overall system mean failure rate. Note that reliability requirements may be expressed as the probability of failure. However, many engineers have a better 'feel' for mean failure rate than probability of failure.

(ii) In the case of maintainability, the design specification should include quantitative requirements for the mean corrective repair time of equipment. A maximum mean corrective repair time can be specified for individual items of equipment or an overall maximum mean corrective repair time for a system can be specified.

The design life of equipment should be stated because this has an important bearing on the selection of components and is essential for calculations if a requirement for the reliability (survival probability) of a piece of equipment is included in the specification.

The equations that are required to calculate that the mean corrective repair time, reliability and the mean failure rate of systems are given in Chapter 3. Some may consider the use of such calculations to be insufficiently accurate since they incorporate estimates pertaining to working environment conditions. However, it is useful to compare their use with stress analysis calculations which most designers readily admit to their work. In stress analysis, assumptions are made with respect to boundary conditions to enable calculations to proceed and knowledge of loadings in practice is generally a first approximation. The reason that few failures occur is that large safety factors are applied to account for the inaccuracy of the mathematical models and loading conditions. How then can stress analysis be said to be more accurate than maintainability and reliability calculations? (The exception to this rule is specialist cases such as pressure vessel analysis where knowledge of loads is more accurately known and complex analytical models, supported by precise experimental methods, are used to determine stresses.)

8.4.2 *Maintenance and operating instructions*

Design specifications should contain a requirement to provide all significant aspects of care and maintenance including, when appropriate:

- operating and maintenance instructions
- lubrication requirements and intervals
- periodic inspections
- adjustments
- safety matters and hazard identification

A design specification may require that a maintenance strategy be recommended for certain equipment or the system.

Therefore all relevant aspects of operation and maintenance can be covered under the terms of the contract.

8.4.3 *Maintenace actions in design specifications*

Maintenance requirements may be specified so that equipment will be fit for its required purposes. In particular, the skill levels and environmental conditions should be accounted for.

The following example gives design requirements that were written for the maintenance actions that were required for a piece of equipment that was to be maintained under special circumstances.

(i) *Minimum skill maintenance methods.* The shape and location of components should be designed to facilitate handling and fitting.
(ii) *Low-force operations.* Damage to protective clothing may result from high contact pressures with tools. Also, excessive heat is generated within sealed clothing when working at a strenous rate. The use of power tools can overcome these problems but a suitable power supply may not always be available, therefore the reliance upon power tools is unacceptable.
(iii) *Uni-directional maintenance.* Procedures and actions should preferably be undertaken from one direction to simplify maintenance and repair operations.
(iv) *Self-draining.* Although systems should be drained prior to maintenance, entrapped fluid will increase the hazard to personnel handling contaminated components. The self-draining criterion must attempt to account for different attitudes of installation.
(v) *Corrosion protection.* Closely tolerated parts and highly stressed components should be protected from seizure.

The short explanation that is given for each point in this specification is far more informative than would be the case if a single list of key words had been given.

8.4.4 Design review
The design specification should state that a design review should be undertaken at different stages in the design project. Chapter 4 discusses design review activities in depth. By specifying that a design review should take place, a specific mechanism is introduced by which the maintainability and reliability may be checked from concept design to detail design.

However, the design specification should be clear about the nature of a design review and all stages and activities should be identified, see Chapter 4, Section 4.3 in particular. A general statement that a design review is required is open to misinterpretation and the design contractor may only expect a final check on drawings to be undertaken. A final check on drawings in all but a very small project is most likely to be a waste of time as changes may not be possible so near the end of the project.

8.4.5 Requirements for the use of specific design methods
A design specification may request that specific design methods be adopted in order to ensure that certain design variables are treated in depth. In the case of maintainability and reliability, the following are examples of those design actions that could be mandated in a design specification.

(i) The use of reliability models to identify, say, the five elements (items of equipment) that most influence the reliability of the system (see Chapter 3, Section 3.6).
(ii) Calculation of the mean failure rates of, say, the five items of equipment that most influence the reliability of the system (see Chapter 3, Section 3.7).
(iii) Undertake a failure mode and maintainability analysis of, say, the five items of equipment that most influence the reliability of the system (see Chapter 7).
(iv) Carry out an equipment evaluation (based on one of the specific methods identified in Chapter 5) for the most significant parts of the system.

Therefore, although no quantitative objectives have been set, the use of the specific design methods has ensured that maintainability and reliability has been considered seriously, quantitatively and effectively.

Consequently, the results of the design analyses may be reviewed and decisions taken concerning the acceptability of certain items of equipment.

8.5 Changes to design specifications

Design specifications are not fixed in tablets of stone. They change. As a design project progresses, the client may revise the requirements, alternative ideas may be suggested by the design team or difficulties may be encountered that require an alternative approach. In a well managed project the client and design team maintain a continuous dialogue throughout the project so that the client is fully appraised of all circumstances and approves of the direction the project is taking.

When difficulties are encountered, the pressure is on to obtain a feasible solution. Cost and timescale requirements are also present. At these points there is the possibility of neglecting the maintainability and reliability aspects of the specification. It may be essential to do this in order to obtain a solution in the required timescale. But, if this has to be done, then it should be done consciously and with the full agreement of the client. The situation must be avoided in which, at a later stage, the client points out that the maintainability and reliability specification is not met in some respect whereas at the time the particular piece of equipment was designed there was extreme pressure just to generate a solution that worked.

Most contractual agreements allow for a limited number of changes to be made. However, it is not good practice towards the end of a design project to raise the importance of reliability and maintainability. The basic maintainability and reliability characteristics will be embedded in the conceptual design and component quality will have been decided by cost considerations. Time will be limited towards the end of the project so that major changes will not be possible, even if an altruistic contractor was willing to undertake substantial work. The right time to consider maintainability and reliability is when the design specification is derived and during design reviews as the project progresses. Minutes of meetings can form part of the design specification and continual reviews with respect to maintainability and reliability are infinetly preferable to 'last minute checks' of the design, which are essentially worthless.

8.6 Demonstrating maintainability and reliability

There is a point in many projects where equipment or systems are evaluated prior to handover to the client in order to demonstrate that the contract has been fulfilled.

The maintenance times of certain pieces of equipment may be demonstrated before handover. Thus, reliance is not placed on calculations to show that the design specification has been met, although calculations should have been used during the design process. Maintainability demonstrations may show that a piece of equipment may be dismantled and reassembled in a certain time, say in the event of failure. Or, certain maintenance tasks may be demonstrated, e.g. replacement of filter cartridges, heat exchanger tube replacement. The advantage of carrying out operations on site is that the effects of other equipment are included in the demonstration: e.g. the close proximity of adjacent machinery, the location of site services.

Reliability performance is difficult to demonstrate in the short term and calculations are required to predict long term performance. However, during commissioning trials the initial running in failures may start to become apparent. If the failure rate does not reduce during commissioning (and during the initial stages of production) then one may raise serious questions about reliability.

8.7 Standards

Reference is sometimes made to standards in design specifications. This approach can have advantages and disadvantages. International and national standards contain guidelines that are based on good practice. They can be very comprehensive but may well deal with generalities, not clients' specific cases. If a standard is referred to in a design specification then the designer is constrained to follow all aspects of the standard even if the design suffers as a consequence. Also, following all the procedures in a standard may be excessively time consuming and inappropriate. However, it may be that particular aspects of the standard are relevant to the design project and if so, then these parts should be identified clearly and not just the whole standard. Standards may be used to identify particular reliability analysis methods.

Maintainability and reliability requirements that will make a beneficial difference to the outcome of a project will often be derived from thinking about the problem and considering all the pertaining circumstances in depth rather than by reference to a standard as a kind of 'catch all' strategy.

8.8 Responsibility for breakdowns in contracts

The supplier of equipment will usually be responsible for repairs during a guarantee period. In large systems, it may be agreed between supplier and user that the cost of repairs will be shared in some proportion, see reference (**1**). After an agreed period, the user would normally be responsible for any repair costs.

It is interesting to consider the trends in repair costs. During the initial Running In period the mean failure rate should fall (see Fig. 3.1, Chapter 3) and the Running In period should correspond roughly with the guarantee period. When the user takes on responsibility for repairs, the mean failure rate ought to be at or near its lowest value and consequently the cost of repairs should be at their lowest point. If the cost of repairs continues to rise early in the life of a piece of equipment or production plant (assuming that future costs are discounted to an initial reference level), then that indicates that there is a reliability problem. The supplier ought to be responsible since the equipment is not fit for its purpose. However, unless the specification contains the relevant clauses concerning reliability then the user probably has little redress on the supplier. Modern production plants have precise management accounting methods that would enable repair costs to be trend monitored. Contractors may also be used which means that costs have to be precisely monitored. Therefore, if the costs can be identified, then the allocation of future cost responsibility allocation to a supplier is feasible.

In the future, production systems may have equipment supplied on a functional basis. That is, rather than purchasing a filtration unit, the supplier would provide 'equipment that will perform the function of the separation of solids from liquids (to some defined specification) for x hours'. This raises interesting questions concerning reliability, maintenance and repair costs, all of which would form part of a contractual agreement with the supplier of the function.

8.9 Management and control of design projects

The management and control of design projects can have a significant influence on maintainability and reliability. There should be adequate time allowed for design effort. This may appear a rather naïve statement but it is true that, too often, unrealistic times are set for design work to be undertaken and the 'time to project completion' is the only goal project engineers appear to understand. Of course, one is not advocating a slack approach to time objectives, but the setting of *realistic* time objectives is important. If unrealistically short times for design work are set then the pressure is on to find a workable solution. The first feasible answer will be chosen, component selection will be rushed, there will be no time for the systematic *evaluation* of equipment and systems. The consequence is that reliability and maintainability will suffer.

Most equipment and production plants have a life cycle of 10–20 years, even longer in some cases. For the sake of the allocation of a relatively little extra design time, the user may have to put up with equipment and systems of inferior design. The life cycle cost will almost certainly be greater than if a reasonable time had been given for design. Often it is the biggest complainers about the quality of design work who are least amenable to allocating adequate time for design.

Project planning methods are well established that use critical path analysis and resource scheduling. It is worth noting that, when difficulties arise during design work, it is often maintainability and reliability that are neglected as time schedules become tight. The onus is on design management to integrate maintainability and reliability with other design parameters. In this way these important variables are less likely to be dropped from consideration in order to meet deadlines than if they are afterthoughts or secondary considerations.

References

(1) Smith, D.J. and **Babb, A.H.** (1973) *Maintainability engineering* (Pitman).

Bibliography

(1) Thompson, G. and **Evans, J.** (1988) On the use of maintainability criteria in contractual design requirements, *Maintenance Management International*, **7**, 85–92.

Chapter 9

Concept Design

9.1 Introduction

Good concept design lays the foundations of a successful design project. The characteristics and limits of equipment performance are determined by the concept. If a design project proceeds on a basis of a poor concept, then there is a limit to what good detail design and in-service changes can do to rescue the situation.

The term 'concept' is used loosely by many engineers with many different meanings. To some, a concept may be an initial idea or even a vague notion. To others, a concept is not meaningful until investigations and calculations have been undertaken. It is therefore necessary to explain what is meant by an engineering design concept in this book.

> A concept defines and describes the principles and engineering features of a system, machine or component which is feasible and which has the potential to fulfil all the essential design requirements.

A concept design exercise is therefore more than the generation of ideas. It is essential that, on completion of the conceptual design study, detail design work can be undertaken that will lead to a satisfactory design. Although many possible detailed solutions may yet have to be explored during the detail design phase, the basic principles on which a good overall feasible solution is based should be clearly established during concept design.

Concept design is the stage in design activities that comes after the requirements have been specified and before the detail design work begins to 'firm up' the design as particular elements are designed or selected. In the design of a product or machine, concept design is the stage that precedes the preparation of the general arrangement drawings which in turn leads to the detail design of particular components.

In the design of large systems, a system concept will be generated after which the design of functional units (which themselves may be large machines) will follow. Concept design exercises may subsequently be carried out as part of the design exercises for the functional units so that there may be several concept design studies carried out within a large project at different levels of design.

Concept design studies necessarily consider many design parameters. Maintainability and reliability are just two of them. When thinking of concept design, one is required to take into account the whole of the design problem and then seek ways to 'inject' maintainability and reliability considerations at the appropriate time. The development of strong concepts provides just such an opportunity. In this chapter the general aspects of concept design will be discussed first, followed by an approach to strong concept development showing how maintainability and reliability considerations may be included.

9.2 General principles

9.2.1 Elements of concept design

There are four essential elements of concept design:

(i) *Ideas*

Options should be considered, hence there is a need to generate ideas. It is poor design practice to work on the first idea that comes to mind.

(ii) *Description and development of ideas*

The ideas need to be described sufficiently well so that they can be understood and compared. The level to which ideas need to be described varies. If many ideas have been generated, then they may well have to be reduced to a manageable number before they can be developed into concepts (the number depends upon the resource and time constraints). An outline description of the operating principle is appropriate to allow a sub-set of ideas to be identified for further development. For the ideas generated, or for the sub-set of chosen ideas, they should be explored and described in more detail to bring out their characteristics. At this stage the ideas are transformed into concepts.

(iii) *Demonstration of feasibility*

The feasibility of ideas should be investigated before making a final evaluation because it is pointless to consider and compare the attributes of concepts in depth when some may be unworkable.

(iv) *Evaluation and choice*

Finally, the alternative concepts should be evaluated realistically and a choice made of the concept that is the best to be taken on to detail design. In concept evaluation, mental agility is required to envisage the likely outcomes of detail design, and knowledge and experience of engineering systems and components in a variety of applications are important. Different designers may well come to different conclusions because of their different experience or expertise.

In certain cases, it is possible to develop two or more concepts in some detail until a favourite emerges. There is no hard and fast rule. In small product design it is even possible to design and make competing designs in order to test which is best. However, on large projects there is usually no opportunity for such luxury. The system design must be fixed before the design of functional elements can proceed. Therefore comparisons of systems must be made without the benefit of hardware. The size and cost of many items of equipment, and the timescale constraints imposed, are mostly so restricting that only one design can be worked on in detail. The objective must be to make the right choices based on decisions made during design. In certain cases, prototype development may be possible to refine the chosen design but reliance on prototype development should be minimized.

9.2.2 *Ideas*

Good ideas are essential. The first ideas are not necessarily the best and ideas generation should be taken very seriously.

Many ideas generating methods are based on group activities, see Appendix 2. However, in many engineering design studies, individuals have to undertake concept design studies but even so it is possible to elicit the help of colleagues without undue disturbance to their work. A route to good ideas generation is shown in Fig. 9.1.

The initial, personal ideas should be recorded as and when they arise.

If possible, a resource group of colleagues should be formed temporarily to carry out an ideas generating exercise on the problem. The problem should be outlined to the members of the group before the meeting. It is important that the ideas generating exercise is carried out on a professional level, otherwise it is a waste of time. For example,

Initial personal ideas + Ideas from a resource group + Further personal ideas = A good set of ideas for evaluation

Fig. 9.1   Ideas generation

brainstorming sessions are very valuable exercises but they become useless if the rules of the brainstorming are not observed rigorously.

An incubation period should then pass to allow more ideas to emerge because the previous ideas generation exercises will act as stimulants to further ideas generation.

Therefore, a good set of ideas can be generated from personal consideration combined with a resource group contribution.

9.2.3 Description of ideas

Ideas need to be described adequately, otherwise they cannot be understood and considered properly. Initial evaluations may be made using the briefest of descriptions of ideas, but as the number of ideas are narrowed down to a handful for deeper consideration, then more in-depth descriptions are required. Some or all the following may variously be used to describe and develop ideas effectively:

- written descriptions
- sketches
- precise geometrical constructions
- engineering drawings
- equations, and
- quantitative data

Different ideas have different description requirements. However, it should be possible for someone other than the idea originator to read the description and understand what the idea is.

Generating a good description of an idea is an important step to formulating the idea. This is not detail design, but rather it is the method by which ideas are thought through and clarified. So-called 'half-baked' ideas are those that are not properly considered and described. Concepts that are not described well could contain serious problems and create serious difficulties should they be chosen for detail design. Written descriptions should be precise and refer to the salient aspects of the design. Reference to other known applications or technology can save much writing.

Sketches should be well made and be informative. Rough sketches containing only an inadequate outline description often indicate poorly thought out ideas. Sketches should contain clearly drawn scrap views, insets etc. to explain important features. In certain cases, engineering drawings may be made to explain ideas if it is convenient to do so. However, in such cases, care should be taken to state clearly that the drawings have been made to explain the concept only. Also, if designers become too involved with detail design work when making engineering drawings at the concept stage, then they may become too committed to a particular design. In some cases, geometrical constructions are needed to explain a concept. For example in mechanism design, the feasibility of the motion needs to be determined using precise constructions in the concept design study and the detail design phase that follows will decide the link and bearing sizes etc. Equations may be used to illustrate principles of operation, e.g. the method of operation of a lifting and manipulation device, illustrating the mechanical advantage gained.

9.2.4 Demonstration of feasibility

Feasibility may be demonstrated by qualitative and/or quantitative methods. Qualitative investigations are often used to demonstrate feasibility. The investigations may use various forms of justification including:

– similar proven applications
– references to other technologies
– the use of well made sketches to show how particular layouts can be achieved, and
– lucid engineering arguments based on sound judgement

Many concepts can be satisfactorily explained and justified solely by good, descriptive engineering arguments.

Quantitative calculations should be carried out, if required, to determine whether it is practicable to pursue a particular design in detail. Assume that a pressurized cylinder of particular capacity and proportions is required as part of a system. A simple, first order calculation should be carried out to determine if it is possible to design a cylinder of the required diameter, which can be made from an economically advantageous material, before proceeding further with the concept. More complex calculations would be performed later, considering attachments etc., during detail design to demonstrate safety.

The identification of the critical parts of the concept that need a feasibility explanation or calculation is as important as the feasibility explanations themselves. A key aspect of the design that is overlooked at the concept stage could cause failure of the whole project if, at the detail design stage, it proves too expensive or impossible to create a feasible design. Such a key aspect could be reliability. If reliability is neglected in concept design then it is sometimes impossible at the later detail stage to create a design with adequate reliability.

9.2.5 *Evaluation and choice*

The objective is to satisfy the design requirements, not to justify personal preferences.

Most evaluations can be carried out by sensible consideration of the advantages and disadvantages of the competing designs. Certainly, in the early stages when numerous ideas have to be considered, engineering judgement should be used.

Systematic evaluation methods should be used with care in concept design studies. Methods that require quantitative estimates to be made with respect to particular criteria are not generally suitable. How can a realistic quantitative assessment of, say, reliability or maintainability be made before the detail design is carried out? The same argument applies to almost all design parameters. Quantified systematic evaluations are useful in design evaluation at the detailed stages when there is sufficient hard data available on which to base judgement. *At all times, methods should be avoided that rely on vague numerical estimates that are 'plucked out of the air'.*

One systematic evaluation method that has proved useful in concept design is described in Chapter 5, Section 5.5. This method works because it is based on comparisons of the attributes of competing proposals rather than on number estimates.

9.3 Principle of strong concepts

A strong concept is one that possesses inherent attributes that will lead to a good quality design with respect to certain essential or desirable features. A weak concept is one that will result in poor design with respect to important requirements, irrespective of the quality of detail design. Of course, a strong concept may well be ruined by subsequent poor detail design.

The advantage of strong concept generation is that it should avoid iteration back to conceptual design activities once significant detail design has been carried out. Although some authors describe the design process as having a possible iteration back to concept design from detail design, such a possibility is impractical in most large design projects and is extremely costly. Iteration back to alter the concept should only take place if there has been some failure in the choice of concept.

9.4 Concept development with respect to maintainability and reliability

9.4.1 Stages of development

Ideas pass through many stages as they are developed into strong concepts. Early on, the concepts may be considered to be immature. As they develop they take on characteristics that determine whether or not they will lead to good detail design. Feasibility studies and other investigations are carried out, but note that detail design work is not being undertaken. The studies are a combination of creative exercises, information retrieval and analysis, and theoretical (and sometimes experimental) work. For example, an exploratory investigation into material properties may be carried out for some critical aspect of the design, or an extensive review of actuator technology could be undertaken to determine if the required positioning accuracy is feasible. Thus, as the concept develops, its main attributes become apparent and the concept becomes ready for detail design.

The initial stages of ideas generation and the formation of immature concepts is basically innovative in character. That is, a wide ranging set of ideas should be generated in order to include as much variety as possible from which advantageous choices and combinations can be made. The development of immature concepts into mature, strong concepts should be different in character. The objective is to strengthen each concept by progressive improvements for it would be unproductive to keep chopping and changing by dropping concepts and starting afresh. Therefore an adaptive creative style is called for in concept development to retain and build on the good features of the initial ideas. (For a discussion of 'Innovative' and 'Adaptive' styles of creativity see Appendix 2.)

The general process of concept development, from initial ideas to a final strong concept is shown in Fig. 9.2.

9.4.2 Adaptive creativity

Beneficial maintainability and reliability characteristics may be introduced by an interative exercise of creative work and evaluation, i.e. divergent–convergent thinking, as the concepts are developed. This divergent–convergent process is repeated until the concepts are ready for final evaluation.

Creative design

Criteria are identified that relate to maintainability and reliability. These will vary from project to project, but typically they will include some of the following (see Chapter 10 for a discussion of criteria):

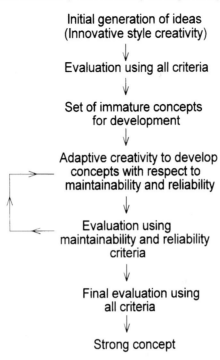

Fig. 9.2   Development of strong concepts from initial ideas

- simplicity and elegance
- minimum number of parts
- modular construction
- accessibility
- sensibly sized components
- ease of adjustments
- minimum number of moving parts
- use of known technology, and
- specific criteria may be used that refer to particular project requirements

Each immature concept is then considered individually and *improvements* made to it using the maintainability and reliability criteria as 'sign posts'.

Concept development may be undertaken by individual designers or by a small group. If undertaken in small groups, the improvements should be made in a constructive atmosphere of 'no criticism', just as in brainstorming. However, the activity should be less charged than in a brainstorming session and brief explanations of the suggested improvements made.

Evaluation
After each concept has been developed significantly, the set may be evaluated with respect to, say, the six most significant maintainability and reliability criteria using the method in Chapter 5, Section 5.5.

Iteration
Further creative development and evaluation may be undertaken to strengthen weak aspects of each concept.

9.5 Final choice

The initial innovative stage generates ideas and leads to a number of immature concepts. The adaptive stage strengthens the concepts with respect to maintainability and reliability.

There is a significant advantage to be gained by developing the immature concepts into mature or strong concepts using maintainability and reliability criteria. The initial ideas generated will be influenced by thoughts of 'will the idea work?' Feasibility studies will have to demonstrate that the ideas will work. By concentrating on maintainability and reliability as the concept takes shape, the resulting concepts will not only be feasible but will also be fit for their required duty.

The final choice of concept should be made with respect to all the salient criteria and not just to maintainability and reliability. Many engineers prefer to limit the number of criteria to (approximately) the six most significant ones. If too many criteria are involved then the evaluation exercise becomes diffuse. Maintainability and reliability should be included in the final evaluation. Judgements with respect to maintainability and reliability will be influenced by prior consideration of the individual, specific maintainability and reliability criteria.

9.6 Brief case study

9.6.1 Problem
A method and equipment was required to decommission a nuclear reactor pressure vessel. The operations have to be carried out remotely and all equipment must be deployed using either a hoist (that provides a vertical lift of 30 kN) or a programmable manipulator (of only 350 N capacity) or a combination of both. The 6.5 m diameter cylindrical pressure vessel is made of steel of thickness 73 mm and has 226 mm of insulation material attached to the outside. The pressure vessel is located within a concrete biological shield with very limited clearance between the vessel and the shield. Even in the confined space there are fixed ladders that were left

after the reactor was constructed. The problem is an interesting one in which a novel solution has to be found. Two very important requirements are reliability and maintainability because the equipment will be used in a radioactive environment with no access for personnel.

9.6.2 Concept development
A short list of immature concepts drawn up from the many ideas generated is shown in Fig. 9.3.

The general criteria for evaluation were:

- hydraulics not preferred
- low powered electrical equipment is good
- minimum waste generation and of a size to fit the standard disposal boxes
- equipment to be deployed by the hoist and/or manipulator
- equipment should fail safe
- reliability takes precedence to high speed operation, and
- repairs should be feasible in the active environment

Experimental feasibility studies were carried out on the use of a torch to cut through the pressure vessel (including the insulation with powder injection) and on the use of an abrasive wheel to cut the insulation and its retaining devices. Design studies of the layout of the insulation, its composition and the methods of attachment of the insulation to the pressure vessel were undertaken. Underside pressure vessel attachments were also considered.

In order to develop a strong concept the following maintainability and reliability criteria were identified:

- simplicity
- robust components
- recoverable in the event of power failure
- known technology
- minimum stoppage time to replace consumable tools, and
- modular construction

The maintainability and reliability criteria were used as signposts during creative design, in an adaptive manner, and decommissioning method '(g) Selective insulation cutting using simple machines' emerged as the recommended method, see Fig 9.3. Figure 9.4 shows the concept developed for just one of the machines, the device to lift up and remove blocks of insulation cut from the outside of the pressure vessel. The drawing is not a detail design nor is it a general arrangement drawing to

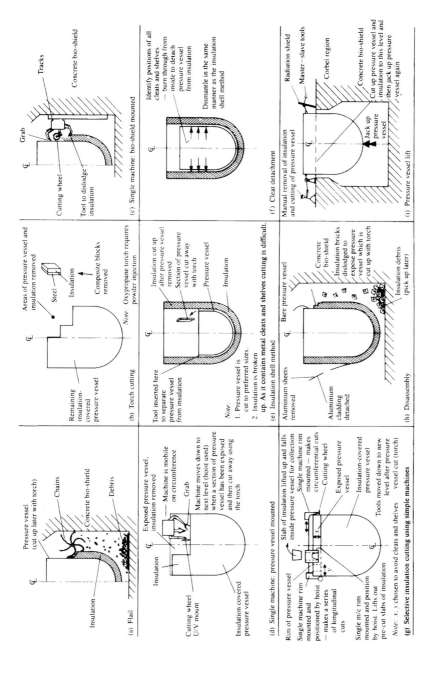

Fig. 9.3 Immature concepts

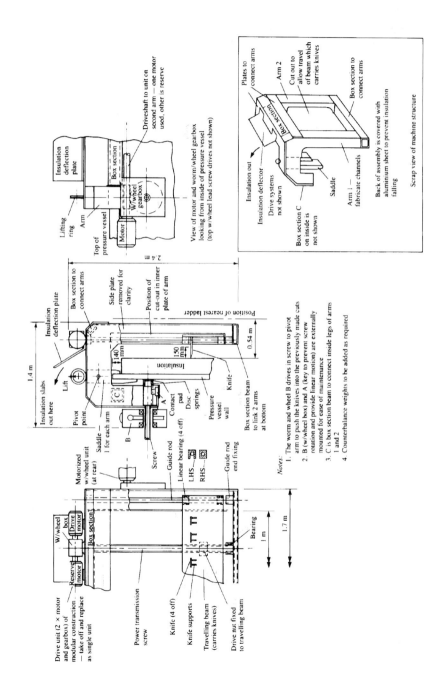

Fig. 9.4 Insulation removal machine

be worked up in detail. The drawing aims to show that a feasible layout of components is possible with particular respect to the space constraints posed by the biological shield. The drawing also contains the thoughts of the designer concerning certain details. For example, the use of worm and wheel gears with two drive motors is given: one motor is a spare and the gearbox and two motors are to be replaced as a single module. As the module is located and fixed in position, the mechanical and electrical connections would be made simultaneously.

Bibliography

Thompson, G. (1992) The design of a method and equipment for decommissioning the WAGR pressure vessel and insulation, *Proc. Instn Mech. Engrs, Part A, J. Power and Energy*, 125–143.

Chapter 10

Equipment Design Principles for Maintainability and Reliability

10.1 Introduction

The objective of this chapter is to describe general principles for the design of equipment with respect to maintainability and reliability. Some examples are then given to show how certain features can be incorporated to achieve good maintainability and reliability. Of course, it is not possible to present a comprehensive portfolio of design features to cover every eventuality. The examples shown are illustrative of what can be done and are intended to stimulate thought for designers' own projects.

10.2 Some general, qualitative guidelines

There are guiding principles that can be used to generate design solutions that are maintainable and reliable. During design work, when ideas are being sketched, when schemes are laid out or when detail design is being undertaken, if the designer is aware of the following principles then high quality design work will be achieved.

Simplicity and elegance
Complicated designs tend to be unreliable and difficult to maintain. Complexity often involves numerous interactions of parts which go out of adjustment more readily than simple designs. Simple, elegant solutions that serve a useful purpose are usually reliable and easy to maintain.

Minimum number of parts
Designs that use many components tend to have high mean failure rates. By reducing the total number of components the overall mean failure rate is reduced. The use of large numbers of parts tends to mean that manufacturing processes become more complex, therefore opportunities are introduced for errors to be made that may well increase the mean failure rate in early life. Also, if there are many components, then there is

more to disassemble and assemble during maintenance, more adjustments to make and consequently the maintainability of equipment with many parts tends to be poor.

A simple design that has a minimum number of parts, each of which is carefully designed to produce an overall elegant solution, is the objective.

Modular construction

Modular construction is probably the single most important design feature that contributes to good maintainability. In the detail design of equipment, it means that components are designed in sets or modules. If a component fails then the complete module of which it is a part is replaced. The mechanical connection between modules should be as simple as possible to make the removal and replacement of modules as easy as possible. If possible, all mechanical, electrical and instrument connections should be disconnected simultaneously in one simple action when the module is removed and, similarly, all connections should be remade when the replacement module is fitted. The principle of modular construction applies to the design of large scale production facilities also.

Large machines should be connected to adjacent equipment using a minimum number of quick release devices. For example quick release hose couplings for hydraulic systems, quick release pipe couplings for process and service pipelines and easy release fasteners for mechanical connections and 'holding down' devices. In the event of a failure in the machine, the machine can then be disconnected quickly from the process or manufacturing line, pulled clear and a replacement machine inserted into the line and reconnected. Mechanical handling gear and a mobile chassis to facilitate machine removal should be specially designed. Of course, such elaborate considerations would only be justified for critical machines in a production process.

Accessibility

Accessibility is significant at several design levels. At the most detailed level, it means, for example providing sufficient space for a spanner to fit over a nut. When machines are designed with numbers of sub-assemblies, then the interaction of the sub-assemblies must be considered to check that one sub-assembly does not adversely affect another. For example maintenance of one sub-assembly should not necessitate the removal of another sub-assembly to provide access. Accessibility also involves the installation of equipment in a production line. A machine may well be designed to be maintainable, but if it is installed in plant in close proximity to other equipment then accessibility to important components may be restricted. For example removal of the tubes from a heat exchanger may be obstructed by other equipment or access to a hydraulic power pack may become difficult because it is 'buried' beneath other equipment.

Sensibly sized components
Components should be designed to facilitate handling. The difficulty in handling small parts that look fine on a drawing is often not appreciated by designers. When working with oily fingers, with tired hands and possibly near the end of a long shift, maintenance tasks that appear straightforward in the design office take on a different complexion in the factory when the pressure is on due to a production line being stopped. Add to these factors the possibility of cold temperatures if outside, or a hot and humid atmosphere in some factory areas (especially in roof spaces), then maintenance tasks that are normally fairly simple become very difficult. Special consideration should be given if protective equipment is worn. Gloves reduce the sense of touch and for health and safety reasons gloves are increasingly being worn.

Adjustments
When certain components are replaced adjustments may need to be made to realign parts of a machine. The set-up procedures used in the factory during the original manufacturing process are not always feasible in the production environment. Positive locations, markings, adjustment screws and clamps etc. should be designed so that realignment after component replacement is reasonably simple. Visibility when working on equipment in a factory environment is not always good. Eye protection is worn in many industrial environments and can reduce vision for intricate tasks, especially in dirty atmospheres when lenses become covered with dust etc. It is easier to locate components against positive 'stops' than to align them visually against a mark.

Moving parts
Parts that are stationary tend to fail less than those that move. If there is relative motion between parts then wear mechanisms are introduced into the machine. Lubrication is usually required for moving parts which in turn raises the possibility of failure if the lubricant is lost, becomes contaminated with solids, or becomes ineffective due to high or low temperatures. Moving parts need to be aligned correctly and are generally more difficult to replace if they fail. Of course, moving parts are essential in many machines but the objective should be to reduce the number of moving parts to a minimum.

10.3 Load and strength

The loads applied to components in service are not constant. They can be described by a probability density function (pdf), see Fig. 10.1. The probability of the load lying between two values is the area under the pdf between the two values. The strength of nominally identical components is also not constant because of variations in their size, the properties of stock materials used in manufacture etc. Similarly, the strength of components can be described by a pdf. The interaction of the strength and load pdfs determines the likelihood of failure.

Consideration of the load and strength pdfs can be used to identify certain actions that can be taken by designers to achieve high reliability when designing equipment.

Load
From the design perspective, there is little one can do to influence the load pdf. However, knowledge of the loads applied is useful because the designer can avoid excessively large components and still achieve high reliability in a cost effective design. Feedback of information from service and analysis of failures provide valuable information for design purposes, see Chapter 13.

Although not under the designer's control, it is worth noting that reliability improves when operators take care of their equipment. If there is a feeling of ownership, for example in the case of production operators maintaining their own equipment, then reliability can improve. This is probably due to the reduction in the high load tail of the load pdf because operators are more careful, do not subject their equipment to high loads and do not abuse equipment. Therefore, machines that have been designed in general with the operator in mind may well prove reliable.

Strength
Improved reliability can be achieved in a cost effective manner by controlling the strength pdf. Precise knowledge of the pdf is not needed, only an appreciation of the principle is required. A safety factor is the ratio of the mean strength to the mean load. If high safety factors are used then the strength and load pdfs will be separated but the use of over-specified components in this way carries a cost penalty. A better way is to use modest safety factors and to control the strength pdf. If the low strength tail of the pdf can be eliminated then reliability will be improved. In effect, what is required is a strength pdf that is narrow, i.e. with a reduced standard deviation. Figure 10.1 shows the cases in which two nominally identical components A and B are subjected to the same load.

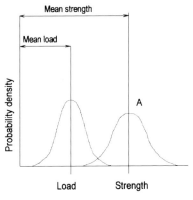 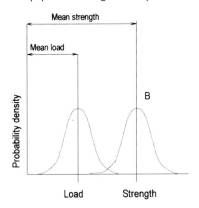

Fig. 10.1 Strength and load PDFs for two components of the same mean strength

The components have the same mean strength but the strength pdf of component B has a lower standard deviation than that of A. Clearly, the load and strength pdfs overlap less in the case of component B so that component B will be more reliable than A.

Note that the safety factors for components A and B are identical because the safety factor is calculated on the mean values of load and strength. However the safety margin of component B is greater than the safety margin for A because the standard deviation of the strength pdf is lower for B than for A. Therefore the use of safety margin is preferred to safety factor when there is sufficient design data.

10.4 Reliability critical dimensions

The low standard deviation in strength can be achieved through design. In component design, the critical dimensions that limit the strength of the component should be identified. These are *reliability critical dimensions*. Design drawings should then specify narrow manufacturing tolerances on *reliability critical dimensions*. Narrow tolerances should not be used generally because that would increase manufacturing cost needlessly.

Figure 10.2 shows a rotating shaft with a central chain sprocket, the ends of the shaft are supported in bearings. The shaft can be considered to be simply supported with a transverse load applied at the chain sprocket at the mid point. A simple analysis of the shaft shows that the central part is subject to high bending loads plus shear stresses due to the transmitted torque. The shaft failed in service due to a fatigue crack that had originated in the fillet radius which was badly machined. The fillet radius is clearly in a critical region of high stress that limits the strength of the

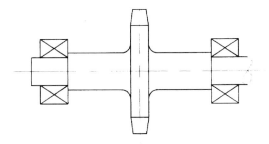

Fig. 10.2 Chain sprocket on shaft

component. In fact the stress level in the shaft was very high at the nominal shaft diameter. Certainly, the fillet radius should have been identified as a reliability critical dimension and the manufacturing quality of the fillet radius should have been checked.

Interestingly, for manufacture and assembly, the salient dimensions are the axial position of the chain sprocket, the axial location of the bearings and the diameters on which the bearings are mounted. The fillet radius discussed above is not significant from a manufacture and assembly viewpoint. Therefore, the dimensional checks for manufacturing with respect to assembly considerations alone will not always include those dimensions that will ensure good reliability.

However, manufacturing quality does not have a role to play in reliability. If components are made within tolerance and if a good inspection system is used to minimize the number of out of tolerance parts, then a significant contribution to reliability improvement will be made provided that specific checks are made on reliability critical dimensions.

Material quality is also important. The supply of raw materials and the quality of heat and surface treatments influence reliability.

10.5 Examples of design

10.5.1 *Fasteners*
Fasteners are very significant features in equipment design because they are a major influence on maintainability. The removal and replacement of most components involve fasteners in one way or another.

Strength of assemblies
Maintainability and reliability considerations are sometimes in conflict. Assume that four bolts of 10 mm diameter are required to hold a

component in place. If the reliability of one bolt is R, then the reliability of the four bolt assembly is R^4 assuming that all four bolts are needed all the time. However, a designer may well choose to use five bolts in order to increase the reliability of the assembly. If the assembly is sound provided that any four of the five bolts survive, then the reliability of the assembly is $R^5 + 5R^4F$ (see Chapter 3, Section 3.6.5). If $R = 0.95$ then the reliability of the assembly increases from 0.815 to 0.977, an increase of 20 percent, if five bolts are used in preference to four bolts. However, the time to remove and replace the five bolts will be 25 percent greater than the time for the four bolt assembly. Therefore, the increase in reliability is made at the expense of maintainability.

One way to avoid this difficulty is to use four larger bolts. Working with nominal dimensions to illustrate the principle, four bolts of diameter 11.2 mm have the same total cross sectional area as five bolts of 10 mm diameter. Consider then an alternative design that uses four over-size bolts of 12 mm diameter. The time to remove and replace a 12 mm diameter bolt will be the same as that of a 10 mm bolt so that the maintainability of the assembly is equal to the one that uses four bolts of 10 mm diameter and better than the five bolt assembly. The four bolts of 12 mm diameter have a larger total cross sectional area and hence strength (for the same material) than the five bolts of 10 mm diameter. Therefore the reliability of the four bolt assembly using over-size bolts of 12 mm diameter should be similar to reliability of the five bolt assembly[1].

Therefore, the best approach is to use a minimum number of large bolts in preference to a large number of smaller diameter bolts in order to improve reliability without adversely affecting maintainability.

Covers and other protection

Covers, screens etc. are very useful design features. They protect equipment from dirt, corrosion and accidental damage and make a significant contribution to improved reliability. However, they can be the cause of high maintenance times if they are held in place by numerous small fasteners that are difficult to handle.

Some simple aspects of detail design can be used to improve maintainability:

---

[1] The reliability calculation depends on more than just total strength. The required mean failure rates of the 10 mm and 12 mm bolts should be calculated. The reliability of the assembly that uses four bolts of 12 mm diameter is P^4 where P is the reliability of a 12 mm bolt. Assuming that all four bolts must always be sound, then $P \geqslant 0.994$ for the reliability of this assembly to be equal to the reliability of the five bolt assembly for which an assumption of $R = 0.95$ was made. For a design life of 50 000 h, the required mean failure rates of the 10 mm and 12 mm diameter bolts are $1.0 \times 10^{-6}\,h^{-1}$ and $0.12 \times 10^{-6}\,h^{-1}$, respectively.

142 Improving maintainability and reliability through design

- Quick release clasps should be used which can be removed more easily than nuts and bolts which have to be unscrewed.
- If screws or bolts must be used, then they should be large and made captive so that they remain attached to cover and do not get lost.

Covers that are difficult to replace will tend to be left off which in turn will mean that reliability will be adversely affected.

Quick release bolts

Figure 10.3 shows a bolt that is designed for ease of maintenance. The head of the bolt is split longitudinally. Each half of the head closes around a shoulder on the bolt, so providing an axial restraint to allow the bolt to tighten. The two halves of the head are retained by a ring 2.5 mm thick located around part of the circumference of the head, see Fig. 10.3(a) and (b). When assembled, the retaining ring is not free to come off the bolt head until the nut has been moved 2.5 mm along the bolt axis. Then the ring will slip clear of the bolt head which can be parted to allow the bolt to be pulled through the bolt hole. In practice, the bolt assembly will not fall apart when loosely assembled. When used in a flange assembly, the bolts can be slackened to separate the faces of the flange to determine if a pipeline has been drained. If there is a need to re-tighten the flange quickly, then this can be done since the bolt assemblies will not fall apart when relaxed. Figure 10.3(c) shows a flange on a 100 mm diameter pipeline that has been assembled with split head bolts. All the bolts in the flange are of the type shown in Fig. 10.3(a) and (b).

Another type of quick release bolt is shown in Fig. 10.4. The bolt has a rectangular head that passes through a slot in one half of the assembly. As the nut is tightened, the bolt rotates through 90 degrees until the head meets the stops which then allows the nut to be tightened. When the nut is unscrewed, the tension in the bolt is released and the bolt will rotate back through 90 degrees until the head meets the stops which align the rectangular head with the slot. The bolt can then be pulled through the slot to allow the assembly to be separated without the need to unscrew the nut completely.

10.5.2 *Access to bearings for condition monitoring*

Designers should be aware of the condition monitoring methods that may be used when equipment is in service, see Appendix 1. The most simple considerations at the design stage can reap benefits later. Figure 10.5 shows two methods of mounting a bearing in the casing of a machine. In method A, a hand-held probe or an accelerometer can be placed readily to the outside diameter of the bearing to monitor the bearing condition. In

Equipment Design Principles 143

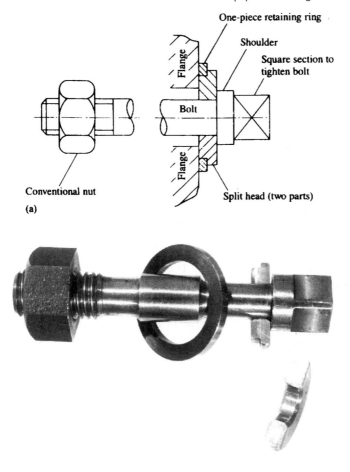

Fig. 10.3(a) and (b)   Split head bolt

method B, which may look more attractive, the outside diameter of the bearing is shielded by the machine casing which makes condition monitoring of the bearing more difficult.

Figure 10.6 shows a pump that has had to be modified to allow an accelerometer to be fitted to monitor the condition of the front bearing. A slot was machined in the casting to provide access to the location of the outside diameter of the bearing. The slot could have been included in the casing during the pump manufacturing process at much lower cost than retro-machining. A simple cover would provide protection from dirt if the condition monitoring facility was not needed.

Fig. 10.3(c) A flange assembly using split head bolts

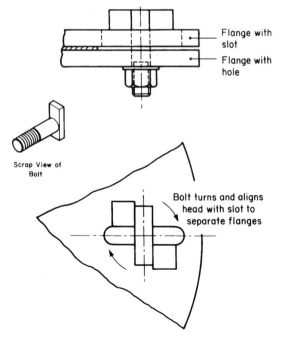

Fig. 10.4 Quick release bolt with a rectangular head

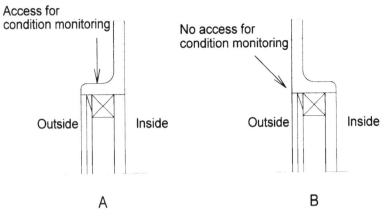

Fig. 10.5 Bearing housing arrangements

10.5.3 *Bearing modules and magnetic drive*

An interesting design problem is to drive equipment in hazardous environments that are shielded by containing all the equipment inside a gas-tight enclosure. The maintenance of equipment inside the containment is very difficult because operations require the use of rubber gloves fixed into the side of the containment wall or the use of manipulators. Special facilities are needed to move spares into and out of the containment without allowing hazardous matter to escape.

Rather than design equipment to be maintained inside the containment using manipulators etc., the problem can be avoided by taking a creative approach. If all electric motors, variable speed drives etc. are outside the hazardous environment then they can be serviced more easily. However, such a solution requires a mechanical drive to be made through the

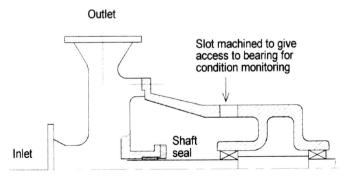

Fig. 10.6 Pump housing modification to allow access for condition monitoring

containment wall. A magnetic drive, see reference (**1**), can be used. The drive consists of two multi-pole permanent magnets placed eitherside of the non-magnetic containment wall. The magnets align themselves so that opposite poles attract. Figure 10.7(a) explains the principle and a 120 mm diameter four-pole ferrous magnet can easily transmit a torque of 1.5 Nm with a 3 mm total gap between the magnet faces. This torque capacity gives a useful power transmission capacity of 0.47 kW at 3000 r/min. Maintenance of the drive is made easy inside the containment by the use of a simple bearing module, see Fig. 10.7(b). If a bearing fails, then the module is simply disconnected from the drive and lifted clear of its housing complete with one half of the magnetic drive. A replacement module is simply dropped into place. Note the simple 'U' shaped saddle

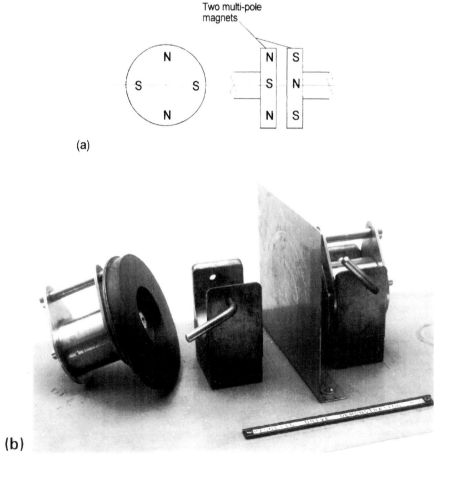

Fig. 10.7(a) Magnetic drive principle; (b) A simple modular bearing arrangement (left hand side dismantled)

location of the bearing module with a simple retaining screw. The saddle incorporates axial location also.

10.5.4 Pipe joints

Pipe joints are found in very many applications including process plant pipe systems, pipework providing building services and connections to a wide variety of machinery. Bolted flanged joints are probably the most common type of pipe joint used and they provide many years of reliable service. However, there are many occasions when a conventional bolted joint is not the best solution to meet particular maintenance needs. There are a number of proprietary pipe couplings available that may well be more appropriate for particular applications from a maintainability perspective. If selected correctly, their reliability should be as good as equivalent bolted joints. Figures 10.8, 10.9, 10.10(a) and 10.10(b) show four different types of coupling and their principle of operation can be understood from the figures. For a further discussion of the reliability of pipe joints and for a more comprehensive review of pipe joints and couplings see reference (**2**).

10.5.5 Valves

Globe valve

Process valves are a common cause of maintenance difficulty, see reference (**3**). Problems are sometimes experienced with stem sealing and internal damage due to particulate matter. The designer faces a dilemma when considering maintainability and reliability. In order to improve

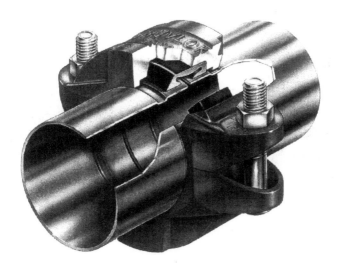

Fig. 10.8 Gruvlock™ coupling (courtesy BSS (UK) Limited)

148 Improving maintainability and reliability through design

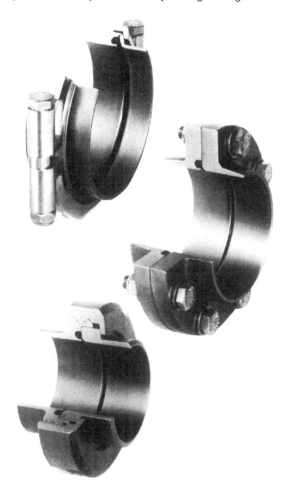

Fig. 10.9 Varivent coupling (courtesy Tuchenhagen Limited)

maintainability, it may be decided to design a valve with a replaceable seat. In the event of seat damage, the seat can then be replaced with the valve *in situ* to avoid having to remove the valve to a workshop to re-machine the seats. However, the introduction of the replaceable seat means that a seal is required between the seat and the valve body which will reduce the reliability of the valve compared with one with an integral seat. The designer must use judgement based on the problems presented in each case to decide whether reliability has to be sacrificed to improve maintainability. A comparative reliability analysis of valves is given in Chapter 5, Section 5.4.2.

Equipment Design Principles   149

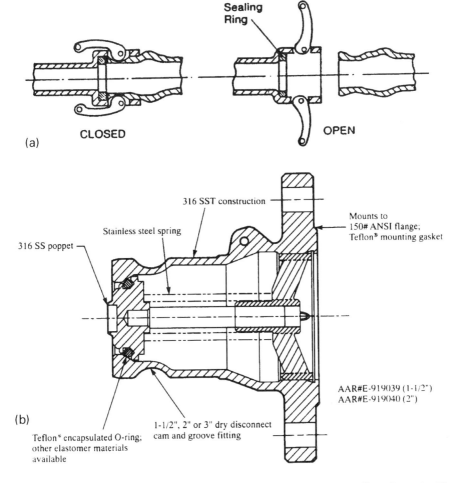

Fig. 10.10   (a) Lever-cam coupling; (b) Lever-cam coupling (one half) and bolted flange combination (courtesy Civacon-OPW). The coupling incorporates an internal shut-off valve to prevent leakage when the coupling is opened

Figure 10.11 shows a globe valve that has been designed for ease of maintenance, see reference (**4**). It has the following features.

– The valve is self-draining when installed either horizontally or vertically. If installed on a slope it is also self-draining provided that the stem is uppermost. Therefore maintenance personnel do not have to worry about trapped fluid.
– All maintenance procedures are undertaken from the bonnet side of the valve and access to the opposite side is not required.

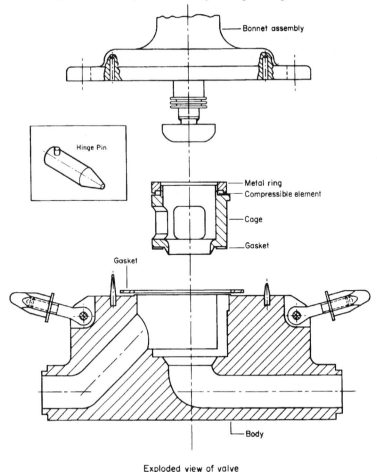

Fig. 10.11 Globe valve design for ease of maintenance

- The bonnet fastenings are all captive to facilitate handling and the nuts on the hinged bolt assemblies are of a reasonable size and shaped to permit easy location by a socket tool. The threads are always covered when fastened to reduce the likelihood of corrosion. Note that when free, the hinged bolts drop away to an angle of 45 degrees so that they are always accessible from above.
- The bonnet is located onto the body by unequal length tapered spigots and is sealed by a compressed fibre gasket.
- The seat is replaceable and is part of a module that comprises: cage, graphite seat gasket and compressible element. To replace the seat, the bonnet is lifted clear, the bonnet gasket is removed and its location surfaces on the bonnet and body are cleaned. The seat module is then

removed, the body is cleaned where the seat gasket lies, and a replacement seat module (with new graphite gasket and compressible element attached) is put in place. The bonnet gasket and bonnet are then replaced and tightened up. The bonnet is tightened sufficiently to compress its fibre gasket and the compressible element ensures that the seat module is simultaneously compressed so that there is sufficient force transmitted to the graphite gasket to make a good seal.

The sequence of maintenance operations can be undertaken using simple tools mounted to the end of shafts if it is required to keep maintenance personnel away from the valve or if the valve is located remotely down a hole.

Gate valve

Figure 10.12 shows the general arrangement of a wedge gate valve that has been designed for ease of maintenance. The valve has two seat modules that are replaced with their body seat rings in place on the module. The bonnet/body connection also uses a similar seat ring instead of a flat gasket to avoid the requirement to control precisely the tightening torque. The valve seats can be readily replaced and Fig. 10.13 shows a view of the valve seat module in position (with the bonnet removed) and a view showing the seat module being lifted out of the valve. The valve

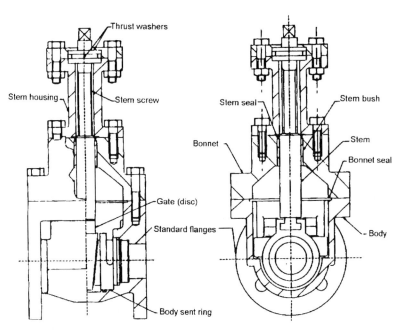

Fig. 10.12 Gate valve design for ease of maintenance

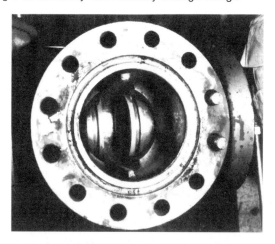

Fig. 10.13 Gate valve maintenance procedure

shown is 100 mm nominal bore and made in steel. It has been subjected to the following pressure test cycle: pressurize to 110 bar with water at ambient temperature, release the pressure, heat to 280°C and cool. No observable leakage occurred in ten complete cycles after which the test was stopped. For further details see reference (**5**).

10.5.6 Multi-function connection

Figure 10.14 shows a multi-way connection that has been designed to connect several hoses and instrument connections simultaneously. The connections are mounted on two plates, each plate contains one half of each hose coupling and electrical connections. The spigots align the two plates when they are brought together. Some designs feature a compliant

Fig. 10.14 Multi-function connection (courtesy Staubli Unimation Ltd.)

feature using springs to help the two halves align when they are brought together and can be locked together manually or by a hydraulic or pneumatic cylinder.

10.5.7 Centrifuge

Figure 10.15 shows a centrifuge for cleaning liquids that have small solid particle contaminants. The centrifuge has been designed for ease of maintenance. It comprises a bowl (1) of 100 mm diameter that is rotated at 20 000 r/min. Contaminated liquid is fed into the bowl and the solids are deposited on the inside of the centrifuge bowl. Clean liquid then sprays out from the top of the bowl and is collected in the catchment container (9) before passing out of the centrifuge on a flexible hose. The bowl is mounted on a shaft (2) by a quick release device (3). The shaft is mounted on two angular contact bearings (4) that are located in a module (5). The module is held in the centrifuge frame (8) by a lever-cam (6). The centrifuge is driven via a magnetic coupling (7) fixed to the end of the shaft. (The drive system to the centrifuge is by an electric motor, electronic speed controller and gearbox culminating in the magnetic coupling. The drive system is not shown.) The clean liquid catchment container (9) is held onto the frame by two quick release toggles. The quick release device that locates the bowl onto the shaft comprises three

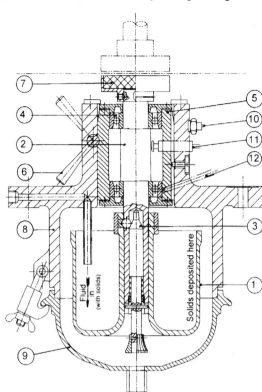

Fig. 10.15 Centrifuge design

captive balls and a spring loaded central rod. The balls are pushed outwards by the rod to provide a positive location in the centrifuge bowl. At speed, the centripetal acceleration generates a high seating pressure for the balls and location does not rely on the spring loaded rod. To release the bowl when the shaft is stationary, the spring loaded rod is pulled down to allow the three balls to unseat from the bowl. The bowl can then be pulled down off the main centrifuge shaft. All the components of the quick release device are 'captive' and the device and shaft remain intact as one unit.

The bearings are light series, sealed for life units using grease lubrication and operate comfortably within their speed range. They are preferred to pressurized air bearings for simplicity, to self-pressurizing air bearings for reliability and to pressurized oil bearings to avoid contaminating the clean liquid with oil.

Maintenance of the centrifuge is simple. The routine maintenance operation is the replacement of a dirty bowl with a clean bowl. Access to the bowl is made by releasing the 'chunky' toggles that hold the catchment container which can then be removed and set aside (the fluid exit is a flexible hose). The bowl and its self-contained quick release device can then be detached from the main centrifuge shaft. An identical clean bowl is quickly attached to the shaft, the catchment container is replaced and the centrifuge is re-started. The dirty bowl is then cleaned and made ready as a replacement for the bowl in service when that becomes 'full'.

Bearing failure and replacement must be considered. If a bearing fails then the whole bearing module comprising two angular contact bearings, the centrifuge shaft and one half of the magnetic coupling is replaced. To remove the module, the catchment container and centrifuge bowl are first removed. The lever-cam (6) is rotated upwards through 90 degrees so that the complete module (5) is free to be pulled down through the centrifuge (the outside diameter of the magnetic coupling is less than the outside diameter of the bearing housing). The module is pulled down through the centrifuge to avoid disturbing the whole centrifuge. In general, it would be more time consuming to disturb the centrifuge than to remove the bowl which has already been designed to be quick release. The bearing module can be refurbished by fitting two new bearings and held as a spare, there is no need to throw away all the good components.

Condition monitoring is by vibration monitoring using an accelerometer (10), a capacitance transducer (11), a temperature probe (12), drive motor current and process fluid monitoring. This comprehensive set of methods ensures that all the principal modes of failure can be detected by two methods. The derivation of the condition monitoring methods is discussed in Chapter 7, Section 7.2.3.

References

(1) **Thompson, G.** and **Cooper, J.M.** (1984) An introduction to the mechanical characteristics of magnetic couplings, *Engng Des.*, **10**, 21–23.
(2) **Thompson, G.** (1998) *An engineer's guide to pipe joints* (Professional Engineering Publications).
(3) **Thompson, G.** and **McKinney, M.** (1989) A survey and analysis of process plant maintainability problems, *Proc. Instn Mech. Engrs, Part E, J. Process Mech. Engng*, **203**, 29–35.

(4) Thompson, G. (1981) The design and maintenance of valves for process plants handling hazardous fluids, IMechE Conference on *Future Developments in Process Plant Technology*, pp. 51–58.

(5) Thompson, G. and **McKinney, M.** (1991) Design study of a gate valve with respect to maintainability, *Engng Des.*, **17**, 26–28.

Bibliography

(1) Thompson, G. and **Xu, Y.** (1989) Design study of a production centrifuge with respect to maintainability and condition monitoring, First International Congress on Condition Monitoring and Diagnostic Engineering Management, COMADEM Birmingham Polytechnic, Birmingham, UK.

Chapter 11

Design for Reliability

11.1 Introduction

Reliability is often considered as a subject that is separate from the 'performance' of a machine or system. Also, reliability analyses tend to be applied at the detail design stage, when components have been defined, in order to demonstrate that the probability or frequency of failures will not be unacceptable. Sensitivity analyses can be undertaken at the concept stage and in system design to determine critical items of equipment before the detail design stage, but generally reliability analysis is applied to specific, detailed design proposals to check a level of acceptability.

However, it is possible to take an alternative approach to the subject of design for reliability to enable engineering designers to:

(i) formulate a conceptual model of what design for reliability means in design synthesis, and
(ii) develop quantitative methods to use the model to assess and develop design proposals to achieve high reliability

In this chapter, the total design problem is conceived as one in which no distinction is made between 'reliability' and 'performance'. The problem is formulated as one in which a set of performance constraints is defined, including functional performance and component failure, and a design solution is found that maximizes the safety margins with respect to the performance constraints.

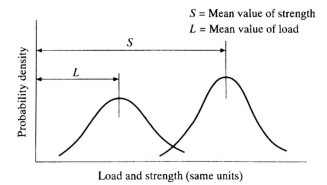

Fig. 11.1   Load and strength probability density functions

11.2 Component strength and applied load

The strength of a component is the ability of the component to sustain the load applied to it. In general, strength may refer to ultimate tensile strength, a fatigue limit, a creep deformation, maximum current, voltage limit or similar quantities. The list is not exhaustive. The relationship between load and strength is commonly illustrated by the relationship between their probability density functions (pdf), see Fig. 11.1.

If the two pdfs overlap, as shown in Fig. 11.1, then there is a probability of failure because of the possible combination of high load and low strength (note that the probability of failure is not the area of overlap). The relative positions of the load and strength pdfs are time-dependent. Various processes will cause deterioration in the strength of components, e.g. wear, corrosion, fatigue and creep. The effect of the reduction in the strength will be to shift the strength pdf towards the load pdf and so increase the probability of failure.

The *safety margin* is defined as:

$$SM = \frac{S - L}{\sqrt{\sigma_S^2 + \sigma_L^2}}$$

where SM = safety margin
S = mean value, strength
L = mean value, load
σ_S = std. deviation, strength
σ_L = std. deviation, load.

The use of a safety margin in design calculations is attractive but it depends upon there being sufficient information available about the strength and load distributions. If the safety margin of a component can be calculated then the reliability of the component can be calculated, see reference (**1**). Even though data may not be available, the appreciation of the safety margin and the need to separate 'load' from 'strength' is fundamental to formulating a design for reliability method.

11.3 Performance and reliability

Reliability is the probability that a component, device or system will continue to perform a specified duty under prescribed environmental conditions for a given time (see Chapter 3). In this approach to design for reliability, reliability includes all aspects of the ability of a product or system to perform. A distinction is sometimes made between the performance of a product or system and its reliability. For example, in a process plant, performance may be measured in terms of output quantities and product purity level and relilability is described in terms of a probability of failure or a mean time to failure of plant items. The distinction between performance and reliability can be unhelpful because it allows some designers to omit reliability from their consideration, leaving the job to others later on.

At the component level, performance and reliability are identical. Consider the case of a bolt. The performance may be measured by its ultimate tensile strength. However, the reliability of the bolt is also determined by its ability to sustain a given load. Similar arguments may be put for other cases to show that there is no difference between reliability and performance at the component level. When a set of components are put together to form a device then they gain a collective identity and are able to perform in a manner that is more than the sum of their parts. For example, a pump is an assembly of components and performs duties that can be measured in terms such as flow rate, pressure, temperature and power consumption. It is the ability of the device to carry out the collective functions that tends to be described as 'performance' whilst reliability is determined by the ability of components to resist failure by some mechanism or other. However, if a pump continues to operate but does not deliver the correct flow rate at the right pressure, then it should be regarded as having failed because it does not fulfil its prescribed duty. It is surely incorrect to describe a pump as 'reliable' if it does not perform the function required of it. One could design a racing car that would be extremely reliable in terms of component failure yet be so heavy and under-powered that it always came in last. The design would be deemed a failure as a racing car.

No distinction will, therefore, be made between performance and reliability in the approach to design for reliability given here. The intention is to design products or systems that fulfil all their required duties. Design considerations may refer to the component level and/or to the collective performance of components. The approach is compatible with the general objective of integrating reliability and maintainability with other activities.

11.4 Equal strength (weakest link) principle

A common sense approach to design for reliability is to create a set of components that are equally strong so that an assembly has no 'weak links'. This is sometimes referred to as 'unity', see reference (**2**). The reliability of a machine or system can be improved by identifying and strengthening the weakest component(s) to that of the nominal level of the other components.

There are two approaches that can be taken:

(i) To specify a minimum mean failure rate and select appropriate components with individual mean failure rates that, when combined, achieve the required reliability.
(ii) To define reliability (including performance) objectives that, when met, achieve an optimum design with respect to overall reliability.

The first method is the conventional approach which is, of course, reasonable provided that dependable failure rates are available or if the component strength and load pdfs are known. However, in many cases none of these are known with confidence but still the designer must try to achieve a high reliability design.

The second method, given in this chapter, is a quantified approach to design for reliability that does not require failure rate data. It seeks to maximize the reliability of the machine or system by ensuring that the system has no 'weak links', whether the weaknesses be component breakages or a failure of the system to perform the required duty.

However, there is another point of view that can be taken with respect to safe design. In some cases, it is better that there is a known 'weakest link' in the system that would enable the onset of failure to be monitored or detected. A system that may be equally likely to fail in a number of different ways, represented by equally high safety margins with respect to all parameters, may not necessarily be the safest system. This design for reliability method is able to take into account a requirement for a system to be stressed to a higher level in a known manner. This will be discussed further in Section 11.5 below.

11.5 Identification of the most reliable solution

An important part of the definition of reliability is the ability to perform within specified limits. Even though precise data may not be available, it is clear from the discussion of strength and load pdfs above, that 'strength' should be separated from 'load' by as much as possible in order to maximize the safety margin with respect to any particular performance criterion.

One can imagine a set of constraints that describe the boundaries of the limits of acceptable performance. If a chosen design solution lies within the space bounded by these constraints then it is a feasible solution, i.e. the solution does not violate a constraint by having an unacceptable performance in any respect. *The most reliable solution is the solution that is the farthest away from all the constraints.* A design that has the highest safety margin with respect to all constraints will be the most reliable. The concept is illustrated in Fig. 11.2 and solution 'A' is the required solution.

If it is required that a 'weak link' be introduced into the system, then the safety margin with respect to the desired failure constraint can be reduced. The optimum solution is then the solution identified as 'B' on Fig. 11.2. The reduced safety margin of solution 'B' with respect to one criterion would mean that the failure would be more likely to occur in this manner. Sensible engineering judgement considering consequences of failure and methods by which the onset of failure could be detected would lead to the correct choice of criterion against which to reduce the safety margin.

During the life of a machine or system there are many events that can cause failure such as: a deterioration in strength with time, a change in the operating environment, an increase in load due to some long term or

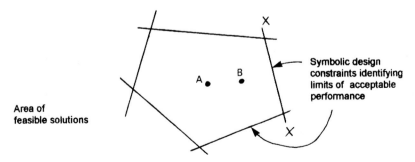

A = Most reliable solution based on the equal strength principle
B = A solution that is allowed to approach constraint X - X

Fig. 11.2 Design constraints and the most reliable solution

transient change. Whatever the cause, failure is due to the coincidence of high load and low strength. The best that a designer can do is to select a solution that has the best safety margin in all salient respects and that solution is the one that is farthest away from the constraints as shown in Fig. 11.2. Thus, the machine or system is best able to handle future coincidences of high load and low strength conditions about which little may be known at the design stage.

11.6 Quantification and measurement against constraints

The objective, then, is to produce a design that has the highest possible safety margin with respect to all constraints. Since constraints will be defined in different units of measurement, and because many different constraints may apply, consideration of a method of measurement is required that will yield non-dimensional performance measures that can be combined in a meaningful way.

In certain cases, for example a pumping duty with a specified pressure-flow rate requirement, the actual performance of a pump can readily be measured with respect to its required duty. In other cases there may be very important criteria that designers would like to maximize or minimize. However open-ended requirements such as 'maximize' are not particularly helpful. It is better to quantify objectives that define limits, for example an objective that is perfectly acceptable and/or one that is completely unacceptable. For example, a percentage improvement on the best performance in the market place could be specified as perfectly acceptable and an existing design used to define objectives below which performance would be unacceptable.

The method of data point generation that is used in parameter profile analysis (see Chapter 6) is adopted here to allow this design for reliability approach to be quantified and lead to an optimization method. Figure 11.3 illustrates how a data point can be generated to measure performance with respect to best and worst limits of performance. The example refers to the stress level in a structure and, importantly, it can be seen how the data point generated draws on the principle of a safety margin. Any stress below $50\,\text{MN/m}^2$ scores 10 since that is considered perfectly acceptable in the judgement of the designer. Any stress level above $200\,\text{MN/m}^2$ scores 0 because that is considered unacceptable.

If there is a case when there are two limits of unacceptable performance, e.g. the maximum and minimum working temperature limits of a valve, then the data point generated depends upon how near the

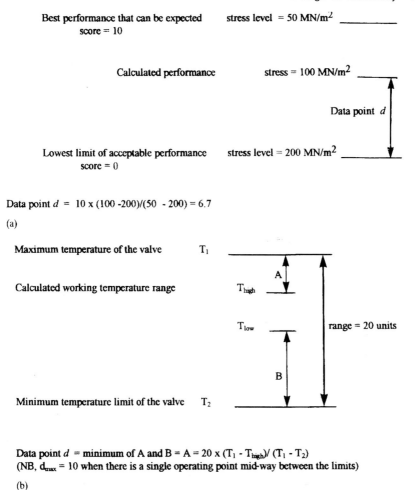

Fig. 11.3 Generation of data points with respect to performance limits

calculated operating temperature range is to the limits of operation of the valve as shown in Chapter 6.

The choice of limits of performance should be made with respect to the application which involves the consequences of failure, knowledge of loading conditions and reliability expectations. If the consequences of failure are high, then the designer can choose to use high safety margins and adopt limits of acceptable performance that are well clear of failure criteria. Similarly, if knowledge of loading conditions is incomplete or if failure criteria are imprecise, then high safety margins can also be used in those particular cases. However, in cases where precise knowledge of loading is known and where confidence can be placed on accurate loading calculations, then acceptable performance limits can be selected at high stress levels so that all components are working near their limits and a high performance product is the result. The designer makes the decision about acceptable performances considering the known information, the consequences of failure and the general engineering circumstances for each application.

If it is required to reduce a safety margin with respect to a particular failure criterion in order to introduce a 'weak link', then the limits of acceptable performance can be modified accordingly. For example, in a stress calculation the lowest limit of acceptable performance may be allowed to approach the yield stress for one component (or a set of components) which would have the effect of permitting higher stresses to be generated in that component than elsewhere.

11.7 Determination of the most reliable design

11.7.1 Basic method

Reliability assessment and improvement may be carried out by the following method.

(i) Identify the criteria against which the design is to be measured.
(ii) Determine the maximum and minimum acceptable limits of performances for each criterion.
(iii) Calculate a set of measurement data points d_i for each criterion following the method described in Fig. 11.3.
(iv) A design proposal that has good reliability will exhibit uniformly high scores of the data points d_i. Any low data point will indicate a performance, perhaps the stress in a component or a system output performance, that is close to an unacceptable limit. Therefore in some significant respect there will exist a low safety margin.
(v) The designer may then review the design proposal in this way and revise the design in an iterative manner to improve low d_i scores. When a uniformly high set of scores has been obtained then the design will be most reliable and will conform to the equal strength principle.

11.7.2 Comparison of designs

If it is required to compare two or more designs, say in an equipment procurement exercise, then an overall rating of reliability may be obtained to compare designs. An overall reliability may be determined by calculating the device performance index (*DPI*):

$$DPI = N \times \left(\sum_{i=1}^{N} 1/d_i \right)^{-1}$$

where N = the sum of the performances considered
       d = the data points generated as described in Fig. 11.3.

The overall *DPI* score lies in the range 0 to 10 and the method of calculation is the same as that described in Chapter 5, Section 5.6.5 for *DPI* when used in equipment evaluation. The inverse combination readily identifies low safety margins unlike addition where almost no safety margin with respect to one criterion may be compensated for by high safety margins elsewhere, which is unacceptable.

Alternative designs can therefore be compared with respect to reliability by comparing their *DPI* scores. The highest score is the most reliable.

Note that caution is required when using any overall rating because simply choosing the highest score may not be the best solution. Each design should always be reviewed to see if weaknesses can be improved upon. Other factors, e.g. cost, novel technology or after sales service, may be the final selection criterion for design proposals with similar overall scores as discussed in Chapter 5.

11.8 Optimization

The process described in Section 11.7.1 above attempts to continually improve reliability in a direction that leads towards an optimal result. If the design problem can be modelled so that it is possible to compute all the d_i scores, then it is possible to optimize mathematically in order to maximize the *DPI* function, as a result of which the d_i scores will achieve a uniformly high score.

An example of this multi-objective optimization has been carried out on antenna structure, see reference (**3**). In the case of an antenna, the design must be optimized for different loading cases since the antenna is used at different angles of elevation. In order to deal with the multiple loading cases, the analysis was developed in the following way. The scores $d_{i,j}$ are determined for each parameter as described in Section 11.6 above, but in this case the analysis is completed for each loading case y_j which depends on the angle of elevation. A matrix of $d_{i,j}$ scores is then constructed based on the parameters considered and the loading cases as shown in Fig. 11.4.

An overall evaluation is then carried out as follows.

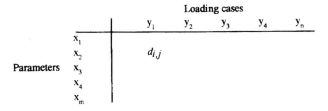

Fig. 11.4 Parameter profile matrix

(i) A parameter profile index (PPI) is calculated for each parameter x_i (analysis of the rows of the matrix):

$$PPI = n \times \left(\sum_{j=1}^{n} 1/d_{i,j} \right)^{-1}$$

where n = number of loading cases.

(ii) Similarly, a loading case performance index (LCPI) is calculated for each loading case y_j (analysis of the columns of the matrix):

$$LCPI = m \times \left(\sum_{i=1}^{m} 1/d_{i,j} \right)^{-1}$$

where m = the number of parameters.

(iii) An overall performance index OPI is then calculated as:

$$OPI = \frac{100}{m.n} \sum_{i=1}^{m} \sum_{j=1}^{n} (PPI)(LCPI)$$

which lies in the range 0 to 100.

(iv) The optimization is carried out to maximize the OPI.

The loading cases were at 0 to 90 degree elevation in 15 degree steps and the antenna was subject to wind loading and self-weight. The parameters used in the optimization were: antenna efficiency, RMS error, maximum displacement, maximum stress, structural mass, structural frequency, -3 dB width, and sidelobe area patterns.

Tables 11.1 and 11.2 show the performances of the original and optimized antenna at the different loading cases. It can be seen that there is a general improvement in the performance of the antenna in all respects, e.g. the maximum stress in the structure is reduced, the efficiency improves, the mass is reduced. Therefore, the antenna will be more reliable than the original design, considering all salient features at component and system performance levels.

11.9 Summary

The method of formulating design for reliability problems described above integrates component failure and functional requirements. It can be used when failure rate data are not available.

The concept of an optimally reliable design solution which is identified as being equidistant from limits of unacceptable performance is an attractive visualization to help in concept design.

The method allows a quantified analysis of a design with respect to reliability to be performed such that weak elements can be improved to achieve the best overall solution and alternative proposals can be compared. Furthermore, if the problem can be modelled then it is possible to use multi-objective mathematical optimization to achieve an optimal solution.

References

(1) **Davidson, J.** and **Hunsley, C.** (1994) *The reliability of mechanical systems* (Mechanical Engineering Publications).
(2) **Pahl, G.** and **Beitz, W.** (1996) *Engineering Design* (Springer-Verlag).
(3) **Liu, J.S.** (1997) *Integrated structural and electromagnetic optimization of large terrestrial space structures*, PhD Thesis, University of Surrey.

Bibliography

(1) **Thompson, G., Liu, J.S.,** and **Hollaway, L.** (1999) An approach to design for reliability, *Proc. Instn Mech. Engrs, Part E, J. Process Mech. Engng*, **213**, 61–67.

Table 11.1 Performances of the original antenna system at seven different working attitudes

Performance parameters	Working/loading cases						
	0°	15°	30°	45°	60°	75°	90°
Antenna efficiency (%)	47.14	47.79	49.63	52.24	54.97	57.05	57.83
R.m.s. error (mm)	0.0557	0.0541	0.0493	0.0419	0.0330	0.0244	0.0203
Maximum displacement (mm)	1.580	1.476	1.272	0.981	0.623	0.275	0.193
Maximum stress (MPa)	18.43	17.96	16.26	13.46	9.741	5.358	4.925
Structural mass (kg)	515.7	515.7	515.7	515.7	515.7	515.7	515.7
Structural frequency (Hz)	9.162	9.162	9.162	9.162	9.162	9.162	9.162
−3 dB width (m deg)	185.1	184.6	182.3	179.2	177.3	175.3	173.6
Side-lobe area in pattern U1	1280	1279	1273	1265	1255	1248	1245
Side-lobe area in pattern U2	1153	1147	1130	1090	1065	1077	1073

Table 11.2 Performances of the optimized antenna system at seven different working attitudes

Performance parameters	Working/loading cases						
	0°	15°	30°	45°	60°	75°	90°
Antenna efficiency (%)	56.12	56.23	56.53	56.95	57.36	57.67	57.78
R.m.s. error (mm)	0.0276	0.0267	0.0239	0.0196	0.0141	0.0077	0.0031
Maximum displacement (mm)	0.773	0.681	0.542	0.407	0.288	0.314	0.328
Maximum stress (MPa)	7.724	7.452	7.659	7.620	8.670	9.129	8.967
Structural mass (kg)	364.4	364.4	364.4	364.4	364.4	364.4	364.4
Structural frequency (Hz)	11.00	11.00	11.00	11.00	11.00	11.00	11.00
−3 dB width (m deg)	177.3	177.1	176.4	175.5	174.5	173.9	173.6
Side-lobe area in pattern U1	1236	1240	1246	1249	1250	1251	1251
Side-lobe area in pattern U2	1049	1048	1081	1078	1079	1074	1073

Chapter 12

Design Actions to Reduce Ongoing Maintenance Costs

12.1 Introduction

There are occasions when designers are called on to reduce the maintenance costs of production systems that are in operation. There may be specific problems that need resolution or there may be a general appreciation that the costs of maintenance are too high and that action is needed. Maintenance costs may be incurred because equipment fails too frequently, equipment may be difficult to repair when failed or equipment may require a lot of attention to keep it running. Whatever the reason, it is quite likely that there are insufficient resources available to adopt the best possible solution to each problem. Therefore, in addition to using expertise to find good solutions, the design manager needs to allocate financial and manpower resources to best effect.

The objective of this chapter is to consider ways in which designers may investigate problems and allocate resources optimally.

12.2 Surveys

12.2.1 Operating records

Operating records can be a valuable source of data which may be used to identify those items of equipment on which maintenance effort is being expended. However, care is required when analysing plant data. If the system downtime has been recorded, then it may well include the time to obtain spares, arrange for personnel to make the repair etc. By redesigning a piece of equipment, or part(s) of it, the designer can influence only the time taken 'on the job' to make a repair and also, possibly, the time taken to identify the cause of the breakdown. If the time taken to effect the repair is a small proportion of the total downtime, then this should be taken fully into account when allocating resources.

12.2.2 Spares usage

The cost of spares and the identification of the machines that are using the spares are both useful indicators of where design actions may be applied to reduce maintenance costs. The costs of spares is an unambiguous maintenance expense which the designer can tackle. The problem may be solved by redesign to reduce loads, wear etc., or alternative materials (or components) may be found that yield a cost effective solution.

12.2.3 Maintenance personnel job records

In some cases, maintenance personnel job records will suggest that certain items of equipment incur a high maintenance expenditure. However, care must be taken when interpreting such data.

For example, in one case it was reported that the maintenance costs associated with the breaking and remaking of bolted flanges was excessively high, based on recorded maintenance times by plant personnel. A call for a new pipe joint was made. On investigation, it was found that the highest proportion of time was spent by fitters standing alongside the flange after completion of the job, waiting for production conditions to be re-established to determine if the joint was leak tight. The time to break and remake the flange was but a small part of the total time recorded by fitters. Interestingly, an alternative joint was recommended for certain cases, but not one that had a quick break and make time. A permanent, welded joint was recommended for those joints which were there only for construction purposes, thus the need for repair actions was eliminated. Permanent, welded joints have high reliability.

12.2.4 Personnel interviews

Interviews with personnel including line workers, supervisors and managers can be held to identify problems. Whilst this can provide useful information, care must be taken when interpreting the results because very often the most recent problem is reported as the most significant. Line workers tend to see the situation very much from within the prevailing circumstances, concentrating on the detail of their personal tasks. Also, different operators and different maintenance teams can have genuinely different experiences. One team may be able to handle equipment well whereas another team may experience trouble with precisely the same equipment. However, operators and maintenance workers do have valuable contributions to make. The investigator must take a detached position and review the information offered objectively.

12.3 Decision making: allocation of resources

12.3.1 Objective
The objective is to undertake a review of equipment maintainability and reliability problems and to achieve a maximum reduction in the maintenance costs of a production system over its remaining life by the use of limited design personnel and financial resources.

Usually, more problems can be found than there are resources available to solve them. There will be limited design personnel available to remedy the problems encountered. The number of personnel available and the number and size of problems determine the timescale in which problems can be solved. Financial resources will also be limited. Therefore alternative low cost options may have to be adopted in some cases rather than preferred solutions.

12.3.2 Costs
Assume that, for a particular problem, a design solution is proposed to reduce the repair time. The cost elements to be considered are:

(i) The initial cost C to make a change to equipment, including the design resource cost and the direct financial cost of the proposed change.
(ii) The maintenance personnel cost savings m per annum as a result of reducing repair time.
(iii) The downtime cost savings per annum d as a result of reducing repair time.
(iv) The spares cost saving per annum s.

If the production system has n years left to operate, then the cost saving of a particular action i will be:

$$lcc_i = (m + d + s) \times \left(\frac{1}{(1+r)} + \frac{1}{(1+r)^2} + \cdots \frac{1}{(1+r)^n} \right) - C$$

(12.1)

where r is the assumed annual interest rate to discount savings in future years to present day values.

Most companies have precise numbers to account for personnel time and the cost of lost production. Therefore, by considering the reduction in repair time the values of m and d may be found. Also, the cost of spares can also be estimated precisely. Note that the time to organize manpower etc. is excluded since design actions do not affect this time.

12.3.3 Design actions and outcomes

When faced with a particular problem, there are several actions that can be followed. Generally, the main options are:

- design a new machine
- redesign part(s) of the machine, and
- buy a new proprietary machine

Of course, these are not an exhaustive set. For example, it may be possible to modify the production process to avoid any changes to equipment and there may well be options to redesign a number of different parts of a machine.

The designer should generate a range of actions that may be followed to solve a problem, and for each action estimate: the cost of the action, the estimated reduction in repair time and any spares cost saving that might be achieved. Therefore, the initial cost C and the expected annual cost savings m, d, and s can be determined.

Also, for each action, the required design personnel resources R should be estimated that would be required to carry out the action.

12.3.4 Resource allocation

Assume that a study of maintenance problems has taken place and that a number of projects P have been identified. For each project, possible design actions A_i have been formulated. For each action, a cost saving lcc_i is calculated using equation (12.1) and the initial cost C_i and the required design personnel resource R_i are known.

Table 12.1 summarizes the available information.

Table 12.1 Actions, outcomes and resource requirements

	Possible actions	Maintenance cost saving	Initial cost of action	Design resource required
Project 1	A_1	lcc_1	C_1	R_1
	A_2	lcc_2	C_2	R_2
	A_3	lcc_3	C_3	R_3
Project 2	A_4	lcc_4	C_4	R_4
	A_5	lcc_5	C_5	R_5
	A_6	lcc_6	C_6	R_6
Project 3	A_7	lcc_7	C_7	etc.
Project 4	A_8	lcc_8	etc.	
	A_9	etc.		
	etc.			

Therefore, a set of actions should be selected such that:

$\Sigma lcc_i = \text{maximum}$

subject to

$\Sigma C_i \leqslant C_o$
$\Sigma R_i \leqslant R_o$

where C_o = maximum financial resource available.
R_o = maximum design personnel resource available.

This decision procedure ensures that maximum benefit, i.e. maximum gain in maintenance cost savings Σlcc_i, is achieved from the available financial and design personnel resources C_o and R_o, respectively. It does not mean that the best option is selected to tackle each problem. A transparent decision making process such as this can be used to justify the allocation of design and financial resources to production and maintenance managers who will, most probably, want the best actions for their own equipment. Also, by taking the decisions based on the cost benefit of alternative actions, the possibility is avoided of a designer introducing his/her 'favourite solutions' which may not be the most cost effective solutions.

12.4 Creative problem solving

There are occasions when there is a general recognition of need to reduce maintenance costs but specific problem areas are not defined with confidence. In such cases, a broad based investigation of production processes and equipment may be carried out to identify problem areas that require solutions.

One approach is to carry out a detailed inventory of items, followed by a failure mode and effect analysis to identify critical items. For the most critical items, a decision to use a particular maintenance technique or to redesign equipment would then be made. Such an approach is really a development within the existing paradigm because it concentrates on the production system as it is found, focuses on detail and develops solutions to fit the circumstances pertaining at the time of the investigation. Although such a method will enjoy a reasonable measure of success, it is not without criticism: the time to investigate the detail is costly and there is no indication of the possibility of success until the investigation has incurred a high level of cost.

An alternative approach is to look beyond the existing paradigm to identify ways of reducing maintenance costs. A flexible 'top down' approach based on creative problem solving (CPS) principles can be taken to identify particular trouble spots and to find innovative solutions. CPS techniques have been used successfully in many business applications and have the potential to investigate maintainability problems. CPS provides a fresh look at the whole scene. It identifies the salient features of problems, generates a wide range of possible solutions and, importantly, gains acceptance for the solutions by the end users. CPS is described in Appendix 2, therefore only a brief outline of its application to investigations to reduce maintenance costs is given here.

The CPS framework may be summarized as:

(i) *Understanding the problem areas*
    The objectives are:
    – to gain a clear understanding of the challenges and opportunities presented
    – to gather data and information about the types of problems encountered, and
    – to gain a focus on particular problems

    For example, this stage of the investigation would establish if a radical, innovative change is required or if a relatively smaller, but still significant, adaptive improvement would be a preferable achievement. The boundaries to the problem are defined, for example, it would be established if the problem should encompass a rethink of the production process. The types of problem encountered would be found by considering the views of operator and maintenance staff, the perceptions of line and senior managers and by all the ways described in Section 12.2 above. At the end of the exercise, a number of principal problem areas will be identified, but they may yet be redefined during the CPS process to obtain solutions.

(ii) *Generating ideas*
    For each problem, a wide variety of ideas are generated as potential solutions. As a consequence of this exercise some problems may well be redefined to generate solutions. The potential solutions should extend beyond the immediate problem. For example, if a machine has a problematic maintenance history then it should not be immediately assumed that an alternative design or maintenance strategy is required for that machine. Perhaps the problem lies elsewhere? For example, if a machine requires constant dismantling

for internal cleaning, then redesign for simpler dismantling may not be the best option; could the product be made into a better state before it reaches the 'troublesome machine' so preventing the need for frequent cleaning? If equipment has poor reliability, could other production factors be changed to improve reliability? The objective is to find the best, most appropriate solution and not to 'jump in' with pre-conceived ideas to improve the design of particular components or to look straightaway to buy a new machine. Thus, problems may be redefined and solutions found for the revised problem definition.

(iii) *Planning for action*

Company employees, from line workers through middle management to senior executives need to be committed to solutions. Line workers should be aware of the reasoning behind any redesign of equipment. If possible they should be given a sense of ownership of the new ideas. In this way, the new or redesigned equipment will be accepted and treated well which in turn will lead to high reliability. If substantial redesign of production equipment is required, then clearly the decision making processes should involve senior executives since they have to sanction the spending.

For further details of the CPS process and the specific CPS tools that may be used to good advantage at each stage of the process see Appendix 2 and reference (**1**).

12.5 Some problems are not designers' problems

Engineering design is an interesting subject. It attracts enthusiasts. Designers are inquisitive and can usually be relied on to think creatively and to look for solutions to problems. However, enthusiasm can tempt one into finding design solutions for all kinds of problems. Whilst certain problems will undoubtedly benefit from a design solution, one should always keep an open mind and allow maintenance management or other solutions to be adopted if they are most cost effective than design solutions. Designers should be mindful not to impose their own pet solutions at every opportunity.

References

(**1**) **Isaksen, S.G., Dorval, K.J.,** and **Treffinger, D.J.** (1994) Toolbox for creative problem solving, Creative Problem Solving Group, USA.

Bibliography

(1) **Thompson, G.** (1982) A decision aid for the allocation of design resources to reduce plant operating costs, *Maintenance Management International*, **3**, 209–221.

Chapter 13

The Feedback of Information to Design

13.1 Introduction

All design activities use data. Designers will regularly look up material properties etc. without a second thought. However, one of the most valuable sources of information, operating and maintenance experience is commonly unused by designers. The reasons for this are various: there may be no mechanism to collect data, if data are collected they may be in a form that is unsuitable for design purposes, designers may think they know what goes on or even operations and maintenance personnel may be unwilling to co-operate in a data collection scheme. Confidentiality considerations may restrict the dissemination of information, e.g. an operating company may not wish to give all its information to design contractors to prevent competitors gaining an advantage. If good data from experience are not available, no matter what the reason, then serious thought should be given to the acquisition of this valuable design information. Data on all salient design parameters are of course useful. Here we are concerned primarily with maintainability and reliability although the principles will apply to most design parameters.

Maintenance information is collected regularly for maintenance management purposes using computer based systems. The information gathered may well not be in a form that is immediately suitable for design purposes, but it should not be ignored. For example, outage times can be indicative of poor maintainability, but care is needed to interpret data because the length of the outage may be caused by the unavailability of spares or manpower shortages which are not equipment design parameters.

Unreliable equipment has a high frequency of failure. To regard this simply as a design fault is to oversimplify the case. The failure may be occasioned by poor care of equipment by operators or even poor maintenance. The manufacturing plant or process may put excessively high loads or production demands on a machine which can lead to failure. A cure for unreliable equipment may be to retrain personnel, change the manufacturing conditions or redesign the equipment. No single answer can be advocated and each case must be considered individually.

Data that affect maintainability and reliability are useful. For example, the loads applied to equipment would assist greatly in the design of new equipment. It may be that equipment failure can be avoided. Equally important may be the possibility to replace over-designed equipment with more economic design in a future design exercise.

Reliability data, in the form of failure rates, are probably the most widely used design data used in the field of maintainability and reliability. The nuclear, offshore and process industries have collected data for many years. The use of reliability data in design is considered specifically in Chapter 3. At this point it is worth noting that reliability data are specific to the operating environment, maintenance skills and manufacturing or process demands. Data from one source are not immediately transferable to another. This is a very important principle that applies to all data collected from operating experience. The use of data in another environment or industry from that where they were collected should be done with the utmost care. It is possible to do so, but careful judgement should be used when comparing the relative conditions.

The objective of this chapter is to consider the use of operating data from a design perspective.

13.2 The use of plant data

Plant records are a valuable source of information for designers. Plant availability records may be consulted to identify those systems and items of equipment that have long downtimes and high frequency of failure. Maintenance records will also show which machines have absorbed most maintenance resources.

There are potential misleading conclusions that may be drawn from the analysis of production data. Recorded downtime may be significantly different to the actual repair time. Consider the events associated with a failure as shown in Fig. 13.1. The system failure is first recorded and there may be a delay until the repair work is begun. The delay may include the time to organize manpower and/or obtain spares. Then follows the actual repair time which may include a significant time to determine the cause of failure which may necessitate a delay whilst spares are obtained. Once the system is operational again and production

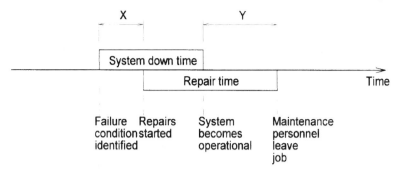

Fig. 13.1 System down time events

restarted, maintenance personnel may need to carry out post start-up checks before they leave the job. The periods X and Y identified in Fig. 13.1 may be of very different magnitudes, therefore system downtime is not always indicative of the repair time. Since the designer can only influence the actual repair time, it is important that this time is identified.

Maintenance records may be used to determine which equipment has long maintenance times. For the designer, a record of the time taken to return a machine to a satisfactory operating condition may, alone, be insufficient. The times to determine the fault, to repair failed components and to make adjustments and set-ups prior to or during initial operation are significant if appropriate redesign actions are to be carried out. Secondary effects can also be misleading. A failure of one item can trigger another failure. Maintenance data may suggest that both items have a poor failure rate since both may well have required attention by maintenance personnel (including adjustments, not just breakages), whereas in reality only the first item has poor reliability from the perspective of initiating failure.

13.3 Data collection

There is a dilemma in data collection. A balance must be struck between obtaining sufficient useful information and requesting so much data collection that an unreasonable burden is put on maintenance personnel. If too much data are requested then the collection system will not be used. Also, care is needed in the level of detail requested. For example, a maintenance fitter should not be expected to identify a fatigue failure. The technology level of data collection must be appropriate to the person generating the data.

Computer based collection systems handle data more easily than human systems, but they can be stupid. For example, spelling mistakes on data entry can generate different failures. The repeated failure of a bearing could be recorded as four separate single events if the data entry is as follows: 'bearing', 'bearin', 'baring', 'bearings'. Instances of all these spellings have been recorded. If possible, a standard system of data entry should be used to avoid spelling variations ruining data analysis.

From the design perspective, it is useful to know how new systems and components are performing, whether good or bad. If failures only are recorded, then good performance is only inferred. The designer may wish to know if particular problems have been experienced or perhaps solved by maintenance personnel. It is not uncommon for problems with troublesome machines or components to be solved by production or maintenance personnel without reference back to design. Designers then carry on in ignorance and specify the same equipment again in the belief that the equipment has performed satisfactorily.

It is not possible to be prescriptive and give precise lists of the data to be collected. Each case deserves individual consideration. The following list gives examples of the types of information that are useful and can be collected.

(i) Name of the item (chosen from a prescribed list to avoid spelling mistakes, see above).
(ii) Cause of failure:
   Example 1 For a valve, a selection may be made from the following:
   – leakage through the valve
   – bonnet gasket leakage
   – stem seal leak
   – stuck open, stuck closed
   – actuator failure
   – broken stem
   Example 2 For a pipe joint, the selection would include:
   – minor gasket leak
   – gasket blow out
   – bolt failure
   – excessive nut/bolt corrosion
   – excessive flange corrosion
   For other equipment, similar lists may be compiled so that the maintenance fitter can select the failure mode.

Example 3 Damage caused by production process:
- out of tolerance parts
- process conditions outside specified limits

Example 4 Failure triggered by another failure, name?

(iii) Corrective action, including:
- time to effect the repair
- spares used
- special skill levels required
- fault finding difficulties

The descriptive answers are useful to designers when carrying out a plant review of the operating history in order to make improvements.

Modern computer software enables more detailed data to be collected precisely and, importantly, without the appearance of an extensive questionnaire. For example, in the above cases of a valve and a pipe joint, the component list would simply include 'valve' and 'pipe joint'. Selection of the component, i.e. 'valve' or 'pipe joint', would then reveal a drop down menu of the types of failure for each case and the fitter would simply select the appropriate mode of failure. There is no need for a fitter to write a description of the failure. Such descriptions should be avoided as more analysis work is required later and descriptions may be imprecise. Records of particular failure modes can be significant in reliability prediction, see Chapter 3, Section 3.7.

13.4 A data feedback system

13.4.1 Principles

The principles of a plant data feedback system that has been specially designed to be an integral part of design activity are shown in Fig. 13.2. The system is designed for designers. It does not have as its sole main objective the compilation of genetic failure rate data that designers may or may not consult. Rather it is an active system that requires input of data from both plant operation and design. It completes a learning cycle for designers. The system has been implemented on a mechanical process manufacturing plant.

The feedback system consists of an active element which receives data from plant operations, a design reference and three forms of output.

13.4.2 The design reference

The Design Reference contains the salient parts of the basic design specification of the plant equipment such as: steam delivery demand rates, production outputs, maintenance times, failure rates etc. It is the

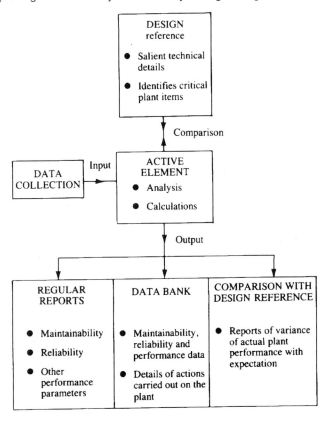

Fig. 13.2   Model of the feedback system

normal data that would be expected in a specification including maintainability and reliability. Not all the parameters in the basic design specification should be included in the Design Reference because the contents of the Design Reference decides what data are collected. There should be just enough data to give useful information but not so much that the data collection process becomes overburdensome. The Design Reference represents the designers' expectations of the plant, it is a set of data that explicitly describes how the designers believe the plant will perform.

Operating data are collected during plant operations and maintenance. When data are collected, they are not only stored in a data bank to build up generic information, but they are also compared to the expectations of performance which have been recorded in the Design Reference. Thus, a learning and monitoring process takes place. Divergence between designers' beliefs concerning plant operation and actuality is avoided. An additional benefit is that, at the design stage, engineers have to consider quantitatively: maintainability, reliability and key aspects of other performance parameters in order to establish the Design Reference.

During the design stage, certain items of equipment will have been known to be near their limit of performance. Also, there may be uncertainty associated with a novel design or perhaps a new supplier is to be used for some components. A competent design team would monitor the subsequent performance of such equipment once the plant is in operation. The feedback system does this by use of the Design Reference in which such items of equipment are identified and particular reports are sent back to design teams to inform them of subsequent performance. Again, the emphasis is on learning from experience.

13.4.3 Output reports

Two kinds of report are generated from the system:

(i) regular reports of performance with respect to maintainability, reliability and other performance data, and
(ii) special reports comparing actual performance with the design reference expectation

The regular reports issued should be of a limited size and their content made relevant to particular design teams. The scope of each report should be determined by each design section head. Thus the amount of paperwork is limited so that engineers do not have to search through lengthy computer printouts to find relevant information, a process that is often avoided by most busy engineers. These written reports should cover those design parameters that were identified as salient features of the specification: the performance of critical items of equipment that are expected to operate near their limit of performance and equipment of a novel kind about which there is uncertainty concerning plant performance. The scope of regular reports may be changed readily in the light of plant operating experience.

A special report should be issued to each design section head when the actual performance of some part of the plant is significantly at variance with expectation. The variance may be due to under- or overperformance; it is particularly important to report this because it is fundamental to the learning process of design. For example, a designer may specify a machine to meet particular performance requirements but in service it proves to be unreliable. Plant personnel may consequently take actions that restrict the use of the machine to a level well below its design specification in order to achieve reliable operations. The result is that the designer believes that the equipment is satisfactory and hence specifies it again. Such cases of divergence between the plant design specification and plant usage are not uncommon.

Therefore, each of the above reporting methods involve designers in the feedback system. They determine the scope of the reporting system and, importantly, there is a learning element where actual equipment performance is compared to expectation.

13.4.4 Database information

The database is built up from plant operating information in a conventional manner, yielding generic failure rate data, maintenance times, other performance data and a history of maintenance actions carried out on plant. Reliability data are sensitive to operating environment; therefore it is advantageous to build up a knowledge base that is relevant to a particular company's operations. While the pooling of data from different sources can be useful, there are obvious limitations.

The history of plant actions is also important. Should there be variations reported between actual plant performance and designers' expectations then investigation will be required. It cannot be presumed that equipment is better than expected if, say, it has a failure rate below that predicted. It may never have been required to perform as originally envisaged. Nor can it be presumed that high failure rates mean that the design is unfit for a purpose. There may be operator error, a supply of raw material may not be within a specification or maintenance routines and procedures may be incorrect.

Provision should be made to interrogate the database to yield the kind of information contained in the regular reports. This provision is made to investigate how one part of a plant performs with respect to another so that designers may review all the equipment (functional units) for which they are responsible and review the performance of sub-assemblies and components that are common to different functional units. Thus, information is made available at different levels: equipment (functional unit), sub-assembly and component. Clearly there is a question of scale here, for a sub-assembly on one functional unit may be a component on another. Common sense is required and the guiding rule is that the lowest level (component) should be that of a replaceable item. Care has to be taken when comparing similar items (for example drive units) on different parts of the plant because they may have different operating duties and environments.

Figure 13.3 shows the interrogation system hierarchy and the screen information that should be made available. The starting point for the interrogation of the database is the reliability model of the plant. This is chosen because it illustrates the significance of the functional unit to the plant of which it is a part. From the reliability model, a functional unit is selected and the screen gives its reliability and maintainability based on its operating history and that of each of its constituent sub-assemblies. A sub-assembly may then be selected to give the information pertinent to its component parts.

Another route into the system should be to list all the equipment (functional units). The mean time between failure, mean time to repair and performance parameters are given alongside the expected values of the variables. This is a direct comparison between actual plant performance and designers' expectations. Experience in the use of such a system as this has shown that this comparison was meaningful and feasible only at the functional unit level. From the comparison screen, it is possible to select a functional unit and review its sub-assemblies and components as described above.

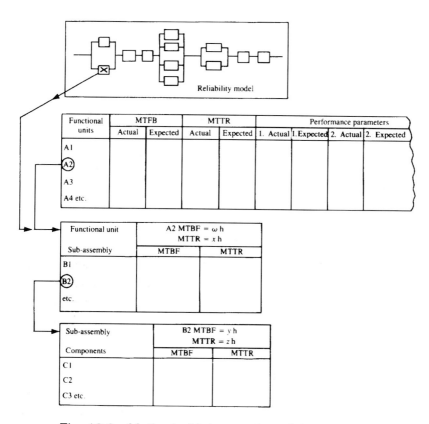

Fig. 13.3 Method of interrogation of the database

13.4.5 *Compiling the Design Reference*

In order to compile the Design Reference, it is required that designers commit themselves to expectations of performance. The experiences of implementation of the feedback system are next given to illustrate some of the problems encountered and how they were overcome.

Discussions with designers have revealed no reluctance to identify salient features of design specifications which refer to performance parameters, e.g. steam consumption for heaters, critical machine speeds etc. However, in some cases, maintainability and reliability gave rise to difficulty.

For maintainability, the view was initially expressed by some engineers that equipment was simply 'maintainable' and quantitative consideration had previously not been given to maintenance times. Similarly, some felt that since equipment had a 'design life' equal to the 'plant life' then failures on the plant were outside the designers' control. The view was expressed that a designer's expectation of a failure rate was somehow an admission of faulty design. However, after careful and considered discussion of maintainability and reliability, their meanings and the causes of poor maintainability, it was found that expectations of plant performance with respect to these variables could be made. Reliability and maintainability were found to be best quantified by failure rate and mean corrective repair time respectively and the corresponding time was recorded on the plant.

It is interesting to note that quite divergent expectations for equipment performance with respect to maintainability and reliability can be found, especially if some designers are fairly inexperienced. Repeat discussions may be required if seriously conflicting expectations are made. Examples may be used to start the process of quantifying expectations. Usually, once designers become 'tuned into' making expectations of maintainability and reliability then an enthusiastic approach follows. Experience has illustrated this point in other areas of design. Once the awareness of the designer to maintainability and reliability has been made, then the parameters are taken seriously.

13.4.6 *Discussion*

The special feature of the feedback system is the establishment of a Design Reference against which plant experience is compared. Designers have to commit themselves to expectations of plant performance in order to compile this reference and thereby become involved in the feedback system which reduces the tendency for data feedback to become a marginal activity which designers may tend to ignore. Comparison of expectation with actual performance and the notification of designers when there is significant variation is an important part of the learning process in design. It has been found that this principle is acceptable to designers and that such a system need not be viewed as an examination process of design activity.

Data input to the system begins at the design stage and continues throughout the life of a plant, involving both designers and maintenance personnel. The data collected on plants are seen to be put to use through carefully designed feedback reports comparing expectation with reality. Therefore, the system is less likely to fall into disuse with maintenance personnel being asked to collect data that are perceived to contribute only to the generation of paperwork that is subsequently little used. A feedback system is only as good as the quality of the data collected; therefore maintenance personnel must believe that their efforts are being put to good use.

Appendix 1

Condition Monitoring

A.1.1 Introduction

Condition monitoring techniques are used to detect the onset of equipment failures so that equipment may be repaired before it breaks. By taking action before failures occur, the following scenarios are avoided:

- unsafe conditions
- difficult repair actions involving broken items
- uncontrolled production shutdown, and
- the initiation of secondary failures

There are many condition monitoring techniques used on modern manufacturing plants. Some identify specific problems, some identify a change to the normal operating conditions. In the latter case, periodic measurements are made at a regular frequency and the method is sometimes referred to as 'trend monitoring'. The objective of this appendix is to give a short overview of condition monitoring techniques so that due account of the methods can be taken in design.

A.1.2 Vibration monitoring

Vibration monitoring is a widely used technique. The simplest approach is to measure the 'overall' level of vibration. Vibration levels are recorded periodically and a change in level, usually an increase, tells the operator that something has changed and a fault may be occurring. Further investigation may be needed to find out just what is wrong. However, if the measurement is taken close to, say, a bearing then the operator will be confident that a problem is developing in the bearing. Probably the most common use of vibration monitoring, especially using hand-held instruments, is to detect failures in rolling element bearings.

The vibration signal may be analysed to reveal its frequency components, either in real time or by recording the signal and analysing it later. Frequency analysis enables the cause of the failure to be identified in many cases. For example, if a shaft rotating at 2000 r/min goes out of balance it would be expected to produce an excitation frequency of 33.3 Hz which should be evident in the frequency spectrum. Certain phenomena may produce particular frequency components. For example, a valve that is leaking internally generates specific frequency components that can be used to identify the leak at low pressures. Other, more complex methods of frequency analyses than frequency analysis are also undertaken.

Simple level measurements are usually taken on hand-held instruments whereas accelerometers (or velocity pick-ups) are used to generate signals for frequency analysis. An accelerometer can use a magnetic base for attachment or, preferably be screwed onto a small stud in a housing. Frequency analysis is usually undertaken between 20 Hz–20 kHz, the range of most accelerometers. Many mechanical vibrations originating from normal machine operation also occur in this range, therefore care is needed when interpreting results to ensure that the phenomenon of interest is the one being monitored.

For design then, access to the outside of bearing should be provided for hand-held instruments. If possible, a flat surface for mounting an accelerometer should also be provided. This can be done at negligible extra cost in a casting process by altering the surface profile just a small amount. If bearing locations can be separated reasonably, then they can be monitored separately.

It is advantageous if other sources of vibration are kept clear of critical equipment to facilitate analysis. In production plant design, poor layout of equipment can severely impede condition monitoring. For example, if critical equipment is located adjacent to certain service items such as compressors, then the signals from the critical equipment may be 'swamped' by the vibrations of the service item.

A.1.3 Acoustic emission

Acoustic emission signals are emitted at high frequencies (up to 1 MHz) by various phenomena, e.g. external leakage from a high pressure pipe joint, minute cracking in stressed composite materials. In practice, the upper frequency monitored is limited and frequency analysis is undertaken on wide frequency bands. The equipment used for acoustic emission requires specialist knowledge to operate.

A.1.4 Displacement transducers

Vibration produces displacement. Therefore, if the amplitude of vibration can be monitored then the vibrating behaviour can be analysed. There are instances when it is not appropriate to attach an accelerometer or a velocity pick-up to a component. In such cases, a displacement transducer can be an attractive option. For example a non-contacting capacitance transducer can monitor continuously the small amplitudes of vibration of rotating shafts. Provision to locate a transducer may be made more cheaply when the equipment is designed than by making modifications to the equipment at a later stage.

A.1.5 Temperature measurement

When components begin to malfunction then they often generate heat. The consequent rise in temperature can be used to identify failing conditions, for example:

(i) tight bearings
(ii) leakage of hot fluids through valves
(iii) blockages in hydraulic systems that cause components and hoses to heat up, and
(iv) electrical components that are overloaded

Temperature measurement may be made by a hand-held probe or by a thermal imaging camera. The latter is particularly useful as a complete system can be surveyed quickly and effectively. In both cases, small temperature differences can be detected; the systems are very sensitive.

Production systems that operate at high temperatures may be monitored by thermal imaging. For example, if a furnace lining becomes thin in one part of the furnace then the consequential higher temperature of the outside surface of that part of the furnace can be detected. Piping systems and vessels that have reduced thickness due to corrosion or erosion will have an increased surface temperature in the thinner section if hot fluids are present.

The location of equipment is important. For example, if surface monitoring of a furnace is expected, then the layout of nearby equipment should be designed to give a clear view of the outside of the furnace. Similarly, hydraulic systems should be in clear view to allow a thermal imaging camera to range over the equipment. Other hot equipment should be located away from the site of interest.

A.1.6 Lubricant monitoring

When wear takes place in bearings and in other contacting parts with relative motion, the wear debris is carried away by the lubrication fluid. Periodic sampling of the lubricant will determine the amount of debris present and hence reveal the extent of wear that has taken place in bearings etc. Lubricants may contain dissolved matter from deterioration processes inside equipment, in which case chemical analysis is required to determine what is happening. The design of equipment should include easy access to lubricant drain plugs and provision to take samples of lubricant should lubricant monitoring be anticipated.

A.1.7 Corrosion monitoring

Corrosion is present on many chemical and other plants and it is often advantageous to monitor the corrosion rates of materials. A strip of metal, identical to the pipe, vessel or whatever is of interest, is held in the process fluid and removed for inspection periodically for examination. The corrosion rate of the specimen can be determined precisely and hence judgements can be made of the corrosion of plant items. Provision for the insertion and removal of corrosion monitoring strips can be made at the design stage.

A.1.8 Electrical parameters

By monitoring electrical parameters, for example electric motor current, faults can be detected in equipment. For example, a bearing that is becoming increasingly tight will require an increase in motor power which can be detected by an increase in the current. Similarly, deterioration in production processes can be identified, e.g. a change in the viscosity of a fluid may well become apparent through a change in the power required for pumps, stirrers etc. 'Sticky' valves that could seize can be detected by the increase in the torque required to operate their actuators. The attractiveness of electrical methods, which can be sophisticated, is that generally they require no mechanical modifications to equipment. Therefore they are particularly suitable for use if, subsequent to equipment installation, a condition monitoring method is required.

A.1.9 Manufacturing and process parameters

Monitoring of the quality of manufacture is also a form of condition monitoring. There are many examples. For example, the wear rate of cutting tools can be deduced from periodic size measurements of machined components and the condition of filters in a process line may be monitored by the pressure drop across the filter (increased pressure drop implies the filter is becoming blocked). Therefore, by making certain quality and process checks, the condition of key components and equipment can be determined and their deterioration in performance monitored. Changes can be made at an advantageous time before the manufactured product goes out of specification.

A.1.10 Crack detection

Crack detection is a form of condition monitoring that is used extensively to check structures, castings and welded joints. There are four principal methods:

- radiographic examination
- magnetic particle crack detection
- liquid/dye penetrant flaw detection, and
- ultrasonic flaw detection

Radiographic examination is the most useful of these non-destructive tests, as it will detect flaws lying wholly within the metal. In applying radiographic examination, the metal is exposed to radiation and any variations in the amount of radiation passing through the thickness of the metal are recorded on a photographic film placed below, say, the weld. Accurate diagnosis of individual markings on radiographs demands considerable experience, particularly of the various types of defect likely to be present in welded joints.

Magnetic particle crack detection can only be used successfully on ferro-magnetic materials such as mild steel. In carrying out this method of crack detection the portion to be examined is first strongly magnetized with an electro-magnet and then fine particles of a magnetic material, such as iron or magnetic iron oxide, are applied to the surface. The presence of a discontinuity is revealed by an accumulation of the powder along its edges. Indications seen as accumulations at surface irregularities may obscure, or be mistaken for, defects when the surface is not sufficiently smooth.

Liquid/dye penetrant flaw detection can only be used to reveal defects which break the surface. The technique involves the application of a penetrating liquid containing a dissolved dye. After this has had time to penetrate into any defects present, the superfluous liquid is removed and the treated surface, after drying, is examined. As the liquid which has entered the defects seeps out, the dye indicates any flaws.

Ultrasonic flaw detection involves transmitting a series of pulses of ultrasonic waves (vibrations similar to sound waves but of much higher frequency, usually 500–5000 kHz) as a narrow beam into the target. On reaching a metal surface such as that of a flaw within the metal, or one of the boundary surfaces of the component, the waves are reflected and return to a suitable receiver. The time required for the return of the echo is a measure of the length of the path traversed by the waves and hence of the distance the reflecting surface is from the receiver. Ultrasonic methods can also be used to detect the thickness of pipes and vessels in which case the reflecting surface is the inside of the pipe or vessel (not a flaw) and has potential for detecting thinning due to erosion or corrosion.

Appendix 2

Creativity and Creative Problem Solving

A.2.1 Introduction

Creativity is widely recognized as being an essential part of design. To many engineers however, creativity is a nebulous concept that rests uneasily in the precise, quantitative engineering world. One person may be perceived to be creative, another not so. Many engineers, quite wrongly, believe that they have little creative talent and in some cases even engineering designers consider that creativity has little relevance to their work.

The objective of this appendix is to explain the basic principles of creativity and creative problem solving in an engineering design context. It will be seen that creative problem solving has a relevance and value that permeates all aspects of engineering design. References are given at the end of the appendix for further study.

A.2.2 Creativity

A.2.2.1 Definitions

There are many published definitions of creativity ranging from the very simple to the highly complex. Two common characteristics of creativity are: *newness or uniqueness* and *value or utility*. Also, the following four strands are significant (reference (**1**)).

The person:	understanding the traits, characteristics or attributes of the creative personality
The process:	describing the stages of thinking that creative people use to invent something new and useful
The product:	the qualities of a product which make it creative

The press: the press is the environment in which the person works which may be conducive to or inhibitive of creativity

The idea of an interaction of the person, process, product and press is used extensively by professionals in the field of creative problem solving to understand creativity.

Therefore, creativity involves novelty (which may be a new idea or a new combination of existing ideas) combined with usefulness. The achievement of a creative solution is a function of: the designer(s); the creative processes used; the artefact or manufacturing system being designed; and the environment in which the designer(s) works.

A.2.2.2 *The designer*

Research has been carried out on the creative individual from different perspectives. Initially, psychologists identified and described the characteristics of highly creative people to try to determine their level of creativity. This approach provided information regarding the cognitive and affective characteristics of people who were generally agreed to be highly creative.

The attributes of creative people may be summarized as (references (**2**) and (**3**)):

tolerance for ambiguity	willingness to surmount obstacles
willingness to grow	moderate risk taking
desire for recognition	willingness for recognition
intrinsic motivation	flexibility
fluency	elaboration
originality	capacity to make order from chaos
openness	curiosity
risk taking	imagination
complexity	independence

The abilities, skills and motivations of people are also important and interrelated factors in creative behaviour (reference (**4**)). Characteristics other than level of ability are significant, e.g. motivation to commit time, energy and effort to creative pursuits and skills in the use of strategies for creating. The *Torrance Tests of Creative Thinking* (reference (**5**)) was one of the earliest known instruments that aimed to measure creativity.

Traditional efforts to gauge the creative person focused on determining the 'creative level' of that person based on a fixed criteria common to highly creative, productive people. However the measurement of the level of creativity is not particularly helpful in engineering design, for it may be counter-productive to give labels that could be interpreted as 'not very creative'. This is especially important because individuals achieve different scores on different instruments so the results may not be accurate.

Kirton (references (**6**) and (**7**)) however, has taken a radically different approach to gauging the creative person by examining one's *style* of creativity rather than the *level* of creativity. Other instruments used in the field to determine approach or style are the Myers-Briggs Type Indicator (MBTI) (reference (**8**)) and Belbin's management of teams (reference (**9**)).

Kirton identifies two preferred cognitive styles of creativity: adaptive and innovative. *Adaptors*, prefer to do things better whereas *innovators* prefer doing things differently. Adaptors will prefer to work within the paradigm, typically they will take an idea and improve upon it. In contrast, innovators will prefer to look to a completely different way of doing things. When faced with a problem innovators will generate lots of ideas quickly, some of which may be 'off the wall'. It is important to emphasize that both adaptors and innovators are capable of generating creative ideas and solutions. Kirton has devised an instrument to measure creative style, the Kirton Adaptor Innovator Inventory (KAI). The KAI score is measured on a continuum from extreme adaptor to extreme innovator. Creative style is not related to creative ability and the distinction between creative style and creative level is illustrated in Fig. A.2.1.

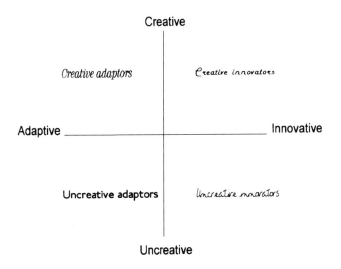

Fig. A.2.1 Creative style and ability

Engineering designers will therefore prefer to be adaptive or innovative in their approach to problems (and of course may have a preferred style anywhere between extreme adaptor and extreme innovator).

The other often-used instrument for determining preference styles in the Myers-Briggs Type Indicator which is based on Carl Jung's theory of psychological types. A person's type is indicated by the individual's combination of each of four different preferences: thinking or perceiving; extroversion or introversion; sensing or feeling; judging or perceiving. The use of these instruments to determine one's style is designed to increase awareness of stylistic differences and their contribution to a given process which in this context is primarily the design process.

A.2.2.3 The creative process

The creative process is concerned with how creativity takes place and is investigated by examining the mental or cognitive processing that occurs as engineers use their creativity. Descriptions of the creative process have been made from the 1920s (reference (**10**)) to the present day. Of the most significant contributions are those by Osborn, Parnes and Guilford. Osborn (reference (**11**)) wrote extensively on the importance of imagination and first introduced the technique of brainstorming which is described later. His model of creative problem solving contains three stages: fact-finding; idea-finding; and solution-finding. Parnes expanded this model by adding a problem finding stage dealing with the necessity of finding the right problem to solve between the fact-finding and idea-finding stage. He also highlighted the importance of solutions getting to the implementation stage by adding a fifth stage, acceptance-finding. The five-stage model of Osborn–Parnes is shown in Fig. A.2.2.

Another significant development, the result of Guilford's work (reference (**12**)) on the structure of intellect, was the addition of so-called 'divergence–convergence' activities in the problem solving

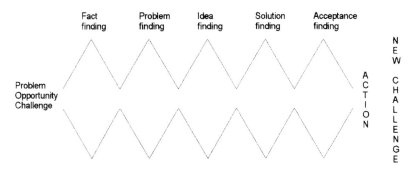

Fig. A.2.2 Osborn–Parnes five-stage creativity process model

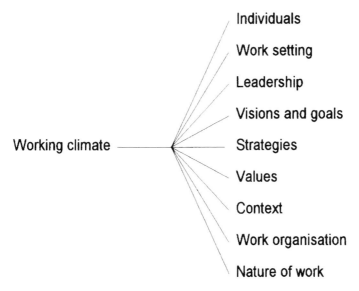

Fig. A.2.3 Characteristic influences of organizational climate (reference (**13**))

process. Divergent thinking is essential for generating a wide range of ideas, and convergent thinking is used to select and evaluate the most useful of the possible solutions. Divergent–convergent thinking is used in each of the stages of the creative process.

A.2.2.4 The product
The product may be a device or system, large or small, with components and/or sub-systems. The creative product, or production system, is one which has novelty and usefulness (value). It may be a radical departure from the norm or it may be an incremental change. Both are creative and both can be of value. The designer must judge whether a radical (innovative) change is required or whether a modification to the existing system (adaptive solution) is required. Incremental improvements have been shown to generate economic gains in production systems.

A.2.2.5 The environment
'Press' is the word used in the Rhodes model to describe the interaction between the person and their situation. It refers to a two-way dependency in which the environment affects the creativity of the person and the person in turn affects the creativity of the environment. The design environment is determined in large part by the organization of and strategy for design within a particular company. Company success depends upon good design. Therefore it is important that companies

understand the nature of design and the characteristics and needs of people who design, in order to create a stimulating design environment. Figure A.2.3 gives characteristic influences that are conducive for innovation and creativity.

The list below shows stimulants and obstacles to a creative environment. In the case of large design project, the size of the undertaking may inhibit creativity because there will be a preference to 'play safe' with so much money involved, i.e. a risk avoiding attitude is taken. The obstacles to creativity include constraints of cost, time and critical evaluation from financial backers. However, in particular design studies which form part of a greater whole, individuals or groups may well feel less restricted and be innovative provided that the environment is conducive and critical evaluation is not an obstacle. It is worth noting that certain creative steps, whether innovative or adaptive in character, may not be apparent because they are 'submerged' in a large project.

Stimulants

    Freedom
    Good project management
    Sufficient resource
    Collaborative atmosphere
    Recognition
    Sufficient time
    Challenge

Obstacles

    Inappropriate rewards, lack of co-operation
    Overly bureaucratic
    Constraint
    Organizational disinterest
    External and critical evaluation
    Insufficient resources
    Time pressure
    Emphasis on status quo

A.2.3 Creativity tools

A.2.3.1 An overview

The creative problem solving process is a comprehensive one that includes problem investigation, ideas generation, evaluation and solution finding and acceptance. A review of all the tools that are available is beyond the scope of this appendix. Methods for evaluation are described in Chapter 5, therefore this review will concentrate on ideas generation, a divergent design activity. There are other significant reasons for paying attention to divergence. 'Brainstorming' is one of the best known techniques, but is one that is subject to most misuse and poor practice therefore it needs to be discussed. Different ideas generating tools are suited to different creative styles, e.g. adaptors prefer different tools to innovators, so different methods are included for ideas generation.

Osborn's brainstorming technique made a breakthrough in applying the psychology of creativity to problem solving activities. Most of the problem solving techniques that have since been developed are based on its underlying philosophy. There are a vast number of creative problem solving techniques, mostly developed in the 1960s and 1970s and applicable to a variety of group problem solving situations. However, many are variations of core techniques and therefore stem from the same basic principles. Here, a review is given of the techniques that are of most relevance to engineers. For a comprehensive review of the creative problem solving process and creative problem solving tools see reference **(14)**.

A.2.3.2 *Brainstorming*

Better ideas are produced when two principles are followed: *deferment of judgement*, and *quantity breeds quality*. Judgement or evaluation of ideas is not permitted during an ideas generation activity since almost any proposed idea can be criticized. All ideas, however ridiculous, are useful, therefore censorship of ideas must be avoided. This encourages a person to come up with more ideas without being fearful of criticism and evaluation. 'Off the wall' ideas have value because they might trigger other solutions. The more ideas generated, the more likely a better solution is arrived at. This emphasis on quantity increases the alternatives. In a brainstorming session, the initial ideas generated reflect known solutions, the next wave of ideas include variations on the existing ideas and some novelty. Effort by the group in the brainstorming session to push for a third wave of ideas may well generate ideas of high quality. Recognition of these 'waves' of ideas by the group facilitator is helpful.

There are four basic rules for brainstorming:

(i) criticism is ruled out (to uphold the principle of deferred judgement)
(ii) freewheeling is welcomed (variety of ideas to stimulate originality)
(iii) quantity is wanted (quantity leads to quality)
(iv) combinations and improvements are sought (listen to others' ideas and improve by additional insights or combination of ideas)

It is important to remember that brainstorming is part of a problem solving process. The group brainstorming activity is one in which an ideas generation resource comes together for the sole purpose of helping the problem holder. Osborn advocated sending group members a memo prior to the meeting outlining the problem so that each person could carry out some thinking before a session.

How different this carefully thought out process is to the oft quoted 'Well, we went away and had a brainstorming session over coffee'. A haphazard activity like this, probably without following the rules for brainstorming, without a facilitator and without being part of an overall creativity process will not be particularly useful and no doubt contributes to the poor reputation of brainstorming in some quarters.

A.2.3.3 Morphological analysis

Morphological analysis is well known in engineering design and is often liked by people with an adaptive creativity style. Possible solutions are generated as follows:

(i) The problem is divided into functions, or even further sub-functions, that must be performed in order to achieve an overall solution.
(ii) Alternative ideas are generated for each function (or sub-function).
(iii) Overall solutions are generated from compatible permutations of the solutions to each function.

The difficulty that the designer faces is to choose the best solution from the large number of options available. In practice, searches may be quickly abandoned once a few acceptable solutions are found.

Functional requirements must be defined with care when using morphological analysis. It is possible to define functions in terms of a preconceived notion of a problem solution. For example, an aircraft could be defined with such functions as wings, tail, tail fin, fuselage, power unit, landing gear etc. This is based on a conventional plane configuration and would exclude a Northrop B-2 which is essentially a flying wing with no tail or fuselage. Function definition should be made such that creativity is not inhibited.

A.2.3.4 Brainwriting

Brainwriting differs from brainstorming in that the generation of ideas is recorded individually on a piece of paper. Some individuals can feel inhibited in group activities. Brainwriting reduces inhibitions because ideas are recorded anonymously and all participants have an equal opportunity to contribute ideas. The technique is practised as follows.

(i) The problem statement is recorded on a worksheet, and below it there are three columns identifying options A, B and C. Each person in the group is given an identical worksheet.
(ii) Each group member writes down three solutions, one under each of the columns A, B and C. If three ideas are not forthcoming then the group member is free to give only one or two.
(iii) On completion of the row of ideas, the group member puts the sheet into a tray and takes a sheet completed previously by another person.
(iv) On the new sheet, the group member writes down up to three more ideas across a row. Each idea is stimulated from the previous idea in the column.
(v) The process is repeated. Each group member may work at their own pace and take worksheets as fast as they like.
(vi) After the session all the ideas are evaluated just as they would following a brainstorming session.

Brainwriting gives opportunities for adaptive and innovative solutions to be forthcoming and is suited to engineering and management problems.

A.2.3.5 Invitational stems, wishful thinking

Using an invitational stem approach for generating ideas can break mindsets effectively. A wishful thinking mode is invited which frees the mind for the generation of ideas without evaluation. Invitational stems can take form of:

(i) Wouldn't it be nice if...
(ii) What I really want to do is...
(iii) I wish...
(iv) How to...

During the initial stages of a creative problem solving, if it becomes apparent that the problem statement needs to be changed from that originally posed, then the use of invitational stems gives an 'up beat', positive climate that is conducive to ideas generation.

Note also that invitational stems can be used in evaluation. For example, rather than commenting that a proposal 'cannot be manufactured quickly', it would be better to say 'wouldn't it be nice if there were fewer parts'. The use of such statements may not appear significant at first but they are a contributing factor to the overall climate, the 'Press' or creative environment in the model for understanding creativity, in which creative engineering will be encouraged.

A.2.4 Discussion

A.2.4.1 Creativity

There are two distinct facets to creativity: *novelty* and *usefulness*. Novelty may take the form of something completely new or it may be a combination of existing ideas or products. For something to be creative it must satisfy a need, it must serve a purpose and make a positive contribution.

Engineering designers are often heard to comment that there is no scope to be creative, especially in the plant industries, and it is said that there is nothing new in certain industries. Such negative attitudes can be dispelled by concentrating on the novelty of original combinations of existing parts and the use of technology from one industry or application to another. Design managers can make a significant contribution by valuing such contributions. The second facet of usefulness is particularly important. Equal emphasis should be given to the usefulness of ideas so that emphasis is taken away from the notion that only people who generate many and often 'wild' ideas are considered to be creative. Here, the outcome of a problem solving exercise is being referred to and not a particular stage when many unusual ideas may be useful as part of an exercise. Therefore, the creative contributions of most design engineers can be recognized, leading to improvements in job satisfaction and performance.

A.2.4.2 Creative style

Creative style is distinct from creative ability. Creative style can be identified on a continuum from extremely adaptive to extremely innovative. Adaptors prefer to make incremental changes, they work within constraints and will seek to improve upon proposals. Innovators prefer to make step changes, they tend to work outside constraints and will look for solutions that differ greatly from existing ideas.

Recognition of creative style is one of the most significant contributions to the field of creativity and it can be measured reliably by the KAI inventory. To many, the innovator is seen as the only creative person, which is to seriously misjudge creative contributions. Innovators and adaptors can both be creative, they can also both be uncreative. The adaptor and innovator, and all styles between the extremes, have a contribution to make. The combination of innovators working in conjunction with adaptors, each recognizing each other's talents, is a very powerful combination in a design team and can improve creativity greatly.

The recognition of creative style has significant implications for design managers. If a design manager does not recognize the importance of style, then the contributions of designers that have a very different style may be undervalued in the team. It may be that designers perceive their contribution to be undervalued so it is important for management to recognize and reward both styles.

A.2.4.3 Invention and innovation

Invention is the design of something that has not existed before and innovation is the process by which new ideas are put into practice. Care is required with terminology here. The word 'innovation' is widely used in this context. However, 'innovator' has a precise and widely accepted meaning with respect to creative style which is not related to the innovation process. No problem exists provided that the words are used in their proper context, but there are significant difficulties when words like innovation are used loosely, which unfortunately is often the case.

With respect to improving creativity, the measure of output from a team should not perhaps be the number of patents it is granted, but rather it should be the number of ideas that have been put into practice. The latter requires innovation. Therefore, to stimulate creativity in a team there is a requirement to put in place processes that will progress ideas to implementation. The process should be supportive and should not be designed so that it is seen as a series of stumbling blocks to be overcome. These stumbling blocks would be obstacles to creativity. The innovation process extends beyond the purely design function to encompass development, marketing, company objectives and finances. Much can be done to stimulate creativity in a design team, but such work can be undone if the work of the team is not valued and integrated within the company as a whole.

A.2.4.4 Designer(s), process, product and environment

High levels of creativity depend upon the combination of the designer, the process used, the product worked upon and the environment in which the work is carried out. It is insufficient to have only the right people or any one of the four elements. It is the combination of people and processes working in the right environment and on products with potential for improvement that will achieve high levels of creativity.

Clearly, design management has a significant role to play here. Creating a supportive atmosphere which is a safe environment is highly conducive to creative design. Designers should be free to speculate, to share ideas and uncertainties without fear of criticism. Looking for problems when faced with new proposals is a sure way to stifle creativity.

Matching people to projects is also important. If a project requires an adaptive solution, say modifications to improve part of a high capital cost plant then it would make sense to give the project to a person with an adaptive style. If a new product is needed to 'break the mould' and make a competitive leap over the competition then the skills of an innovator are called for. Between these extremes, the design manager should bring together adaptors and innovators to best effect according to the project. The creative problem solving process, see Fig. A.2.2, has much to commend it to engineering design. It can be readily seen how the process is applicable on a large scale and on a small scale to find solutions to particular problems. Of great importance is discipline in the use of judgemental thinking. Awareness of where one is in the process and the ability to suspend judgemental thinking when necessary is essential to arrive at a good solution. Creativity is not just about generating ideas. It is about finding solutions to problems.

References

(1) **Rhodes, M.** (1961) An analysis of creativity, *Phi Delta Kappan*, **42**.
(2) **Sternberg, R.J.** (editor) (1988) *The nature of creativity: contemporary psychological perspectives* (Cambridge University Press).
(3) **Isaksen, S.G.** (1987) Concepts of creativity, In Colemont, P., Groholt, P., Rickards, T., and Smeekes, H. (editors), *Creativity and innovation: towards a european network* (Kluwer Academic Publishers).
(4) **Torrance, E.P.** (1979) *The search for satori and creativity* (New York: Bearly Limited).
(5) **Torrance, E.P.** and **Ball, O.E.** (1984) *Torrance tests of creative thinking: streamlined manual, figural A and B* (Scholastic Testing Services, Basenville, Illinois).
(6) **Kirton, M.J.** (1976) Adaptors and innovators: a description and measure, *J. Appl. Psychology*, **61**.
(7) **Kirton, M.J.** (1989) *Adaptors and innovators: styles of creativity and problem solving* (Routledge).
(8) **Myers, I.B.** and **Myers, P.B.** (1980) *MBTI: gifts differing* (Consulting Psychologists Press, Inc.).
(9) **Belbin, R.M.** (1981) *Management teams: why they succeed or fail?* (Butterworth-Heinemann).
(10) **Wallas, G.** (1926) *The art of thought* (Cape).

(11) Osborn, A.F. (1963) *Applied imagination: principles and procedures of creative problem solving* (3rd edition) first published in 1953 (Scribner's Sons).
(12) Guilford, J.P. (1956) The structure of intellect, *Psychological Bulletin*, **53**.
(13) Ekvall, G. (1987) The climate metaphor in organization theory. In B.M. Bass and P.J.D. Drenth (editors), *Advances in organizational psychology: an international review* (Sage Publications).
(14) Isaksen, S.G., Dorval, K.J., and **Treffinger, D.J.** (1994) *Toolbox for creative problem solving* (Creative Problem Solving Group).

Bibliography

(1) Parnes, S.J. (1967) *Creative behavior guidebook* (Charles Scribners Sons).
(2) Thompson, G. and **Lordan, M.** (1999) *A review of creativity principles applied to engineering design, J. Process Mech Engng, Proc. Instn Mech. Engrs, Part E, J. Process Mech. Engng*, **213**, 17–31. Appendix 2 is based on this paper.

Appendix 3

Mean Failure Rate Data

The use of failure rate data is discussed in Chapter 3, Section 3.7. The data given below are suitable for making estimates and should be considered as nominal data. Stress and environment factors should be applied as appropriate. For safety critical applications and where accurate estimates are required, then data should be used that have been collected from similar applications, see Table 3.4 in Chapter 3 for data sources.

Component	Mean failure rate $\lambda/10^6$ h
accelerometers	2.8
accumulator	7.2
actuator	5.1
alternators	0.7
bearings	0.5
bearings, ball	
heavy duty	1.8
low speed, light duty	0.875
sleeve	0.5
roller	0.5
brackets, mounting	0.0125
cams	0.002
clutches	0.04
clutches, magnetic	0.6
couplings	
flexible	0.69
rigid	0.025
covers, protective	0.038
cylinders	
hydraulic	0.008
pneumatic	0.004
drives	
belt	3.9

direct	0.4
filters, mechanical	0.3
gears	
helical	0.05
spur	2.2
heat exchangers	15.0
hoses, pressure	3.9
housings, cast machine bearing surface	0.4
motors	
electrical	0.3
hydraulic	4.3
servo	0.23
stepper	0.37
pins, guide	1.625
pistons, hydraulic	0.2
pumps	13.5
hydraulic drive	14.0
pneumatically driven	14.7
vacuum	9.0
regulators, flow and pressure	2.14
seals	
rotating	0.7
sliding	0.3
springs, simple return force	0.12
tanks	0.15
valves	
ball	4.6
butterfly	3.4
check	5.0
control	8.5
relief	5.7
shut off	6.5

These data are from Rothbart, H. (1964) *Mechanical Design Systems Handbook*, McGraw-Hill and are reproduced with permission of the McGraw-Hill Companies.

Index

Acceptance criteria 42
Accessibility 52, 56, 128, 136
Acoustic emission 192
Active redundancy 27
Adaptors 199, 200, 203, 206
Adjustments 19, 52, 128, 137
Allocation of resources 173
Applied load 158
Assessment criteria 61, 63, 73
Availability 21, 22, 41, 42, 112

Brainstorming 203, 204
Brainwriting 204, 205
Breakdowns in contracts 118

Centrifuge 153–155
Check list 45, 46, 52–54
Convergent design 14
Comparative reliability, 46, 57, 148
Component 1, 8, 121, 135
   analysis 40, 41, 45, 46
   count method 27, 30, 35
   failure 157, 169
   quality 55, 57
   strength 158
Concept 2, 7, 8, 12, 60, 61, 122, 124, 125, 126, 129
   design 1, 6, 7, 11, 12, 14, 121, 122
   development 127, 128, 130
   evaluation 44
   stage 11, 51
Condition monitoring 45, 53, 56, 57, 94, 142, 145, 155, 191, 192
Consequence of failure 96, 98, 99, 104, 107, 163
Constraints 109, 110
Construction 53
Contents of a specification 113
Contracts 109
   breakdowns in 118
Contractual agreements 3, 116
Contractual dispute 111
Control of design projects 118
Convergent thinking 12
Corrective maintenance time 36
Corrective repair times 20, 35, 39
Corrosion 53
   monitoring 194

Cost 53, 54, 112, 173
   saving 173, 174
Covers 141
Crack detection 195
Creative design 127
Creative level 198
Creative problem solving 60, 175, 176, 177, 197, 198, 200, 202, 203, 205
Creative process 200, 201
Creative style 199, 206
Creativity 5, 127, 197, 202, 206–208
   tools 202
Criticality 99

Data:
   collection 181
   feedback 188
   feedback system 183
Database 186
Database information 186
Decision making 5, 173
Decommissioning 3, 49
Design:
   audit 49
   constraints 161
   contractors 1
   evaluation 51
   for reliability 157
   life 23, 60
   models 9
   objectives 110, 111
   phase 5
   projects, control of 118
   reference 183–185, 188
   requirements 40, 121, 126
   review 2, 26, 39, 40, 43, 47, 49, 115
      management and control 47
      team 46
   procedure 40, 43
   specifications 40, 41, 109–112, 114–116
Detail design 1, 6, 7, 14, 56, 125, 157
Device parameter index 85
Device performance index (DPI) 44, 65, 69, 70, 78, 77, 84, 88, 165
Dismantling 19
Displacement transducers 193
Divergence–convergence activities 200

214 Index

Divergent thinking 12, 14
Divergent–convergent thinking 201
Downtime 3
Downtime cost saving 173

Electrical parameters 194
Elegance 128, 135
Environment 3
Environmental conditions 41
Environmental factors 30, 36, 75, 211
Equipment:
  calculations 46
  design 8, 9
  design features 56, 57
  design principles 135
  evaluation 40, 41, 44, 51, 54, 115, 165
Ergonomics 53
Evaluation 51, 52, 60, 118, 122, 126, 129
  of design concepts 60
  of equipment 62

Failure 17, 18, 117, 136, 155, 158, 180–182, 191, 192
  consequences of 96, 98, 99, 104, 107, 163
  criterion 18
  mode 20, 56, 91–94, 96, 98–100, 104, 107, 115, 182, 183
  mode analysis 44, 45, 91, 108
  mode and effect analysis (FMEA) 98, 99, 104, 107
  mode and maintenance analysis (FMMA) 91, 94
  mode effect and criticality analysis (FMECA) 98, 100
  rate 21–23, 33, 61, 160, 181, 183
  data 30, 75
Fasteners 52, 140
Fault 53
Fault finding 19
Fault tree analysis (FTA) 42, 104–107
Fault trees 104, 106
Feasibility, demonstration of 122, 125
Feedback 13, 14, 47, 74, 138, 179, 183
  system 183–185
Frequency 97
Frequency of failure 21
Full active redundancy 34
Functional definition 204
Functional requirements 110, 167
Functional unit 8–10, 121, 187
Functional unit evaluation 40, 41

Gate valve 151
Globe valve 147, 150

Hazan 97
Hazard 97, 106, 114
  analysis 97
  and operability (HAZOP) 106
  rate 22, 23
Hoses 151

Ideas 123, 124
Initial cost 54, 174
Innovator 199, 200, 202, 203, 206, 207
Invention 207
Inventiveness 5
Invitational stems 205
Iteration in design 11

Judgmental thinking 14

Large-scale projects 9
Levels of design 8, 39
Life cycle cost 2, 3, 54
Load 138
Lubricant monitoring 194

Machine analysis 75
Maintainability 1–3, 6, 7, 9, 10, 13, 14, 17, 19, 30, 39, 41, 43, 44, 46, 47, 51, 52, 54, 56, 61, 62, 67, 70, 78, 81–85, 87, 89, 91, 92, 96, 108–113, 115, 118, 122, 126–130, 135, 136, 140, 141, 147, 148, 160, 176, 179, 180, 185, 187, 188
  analysis 45, 115
  prediction 19, 27, 33
  requirements 117
Maintenance 3, 8, 17, 47, 52, 56, 91, 93, 109, 111, 114, 136, 137, 145–147, 149, 151, 154, 171–174, 179
  actions 114
  costs 176
    ongoing 171
  cost saving 174, 175
  information 179
  instructions 114
  management 1
  personnel 181, 189
  personnel job records 172
  record 181
  skill 41, 180
Management 109, 118

Mean corrective maintenance time 21, 67
Mean corrective repair time (MCRT) 19, 69, 73–75, 77, 78, 83, 86, 87, 89, 112, 113
Mean failure rate 23, 24, 30, 36, 60, 75, 113, 115, 118, 135, 160, 211
Mean repair time 39, 111
Mean time:
  between failures 26, 187
  to failure (MTTF) 21, 25–28, 33, 36, 39, 67, 69, 73–75, 77
  to repair 19, 21, 187
Measurement 63
Modes of failure 155
Modular construction 41, 56, 62, 128, 136
Morphological analysis 204
Morphological chart 94, 95
Multi-function connection 151
Multi-objective optimization 165
Multi-skill working 41

Nature of design activity 12
New technology 45, 62
Nominal failure rate 75
Novel designs 45
Novel technology 55, 165

Operating environment 56, 57, 180
Operating instructions 114
Operating records 171
Optimization 5, 165, 166, 167
Overall value 65

Parallel elements 27, 33, 34
Parameter performance index 84
Parameter profile:
  analysis 43, 45, 81, 87, 88, 161
  index (PPI) 88, 166
  matrix 84, 85
Parameters:
  manufacturing 194
  process 194
Partial active redundancy 29
Performance:
  constraints 157
  criteria 62, 161
  data 185
  parameter index 85
  variables 39
Personnel 53
  interviews 172
Pipe joints 147
Plant data 180

Prediction of reliability 25, 30, 33, 183
Probability density function (pdf) 138, 158
Probability of failure 20, 21, 22, 25, 27
Product design 9
Production availability 22

Quality control 55
Quantitative requirements 113
Quick release bolts 142

Re-assembly 19
Redundancy, active 27, 28
Redundancy, standby 28
Reliability 1–3, 6–10, 13, 14, 17, 18, 21–23, 26, 30, 33–35, 39, 41, 43, 44, 46, 47, 51, 52, 54, 56–58, 61, 62, 70, 78, 81, 89, 91, 96, 108, 111–113, 115–118, 122, 125, 127–130, 135, 138, 140, 141, 147, 148, 154, 157, 159–161, 164, 167, 177, 179–181, 185–188
  analyses 43
  assessment 164
  comparative 46, 57, 148
  criteria 62
  critical dimensions 138, 140
  model 43, 58, 115, 186, 187
  modelling 26, 57
  objectives 110
  prediction 25, 30, 33, 183
  requirements 109, 117
Repair 93
  actions 91, 93
  time 21, 33, 61, 180, 181
Requirements 6, 109, 112
Requirement in a specification 6
Resource allocation 174
Review team 47
Risk 1, 96, 97
  assessment 96
Robust 61
Running-in 24, 118

Safety 1, 39, 47, 97, 114
  assessment 17, 97, 99
  factors 138
  margin 157–159, 161, 163, 164
Scale factors 74
Sensitivity analysis 42, 157
Series elements 26
Servicing 55, 57
Simplicity 53, 62, 128, 135
Spares 53, 55, 57

cost saving 173
usage 172
Specifications 1, 6, 11, 42, 109
Standards 117
Standby redundancy 28
Strength 138
Stress factors 30, 36, 75, 77, 211
Strong concepts 7, 61, 126
Sub-assembly 8
Sub-system 8
System 8
   analysis 70
   concepts 9
   design 11
   downtime events 181
   evaluation 70
   failure 36, 42, 180
   levels 1
   reliability 28, 34
   requirements 9
   review 40, 42
Systematic evaluation 126
Systematic quantitative equipment evaluation 62

Systems evaluations 43
Systems level 41
   review 41

Technological uncertainty 62
Technology 53, 55
Temperature measurement 193
Terotechnology 2
Trend monitoring 191

Uncertainty 5
Useful life 24
Utility functions 59, 64, 67, 73, 74

Value 5, 63
   judgements 12, 63, 64, 66, 74
Valve 147–149
   gate 151
   globe 151
Vibration monitoring 191

Weak link 161, 164
Wear out 24

TA 174 .T47 1999
Thompson, G.
Improving maintainability